D0128810

More Praise for *Ecoliterate*

"One of the most urgent issues facing humanity is fixing our broken relationship with the earth, on which all life depends. To do that, we have to think, feel, and act differently. With vivid examples and lucid analysis, this powerful and persuasive book shows just how much inspired educators and students can achieve together. It should enlighten and invigorate schools and communities everywhere."

—**Sir Ken Robinson**, creativity expert and author of *The Element: How Finding Your Passion Changes Everything*

"The worlds of thought and action have been fundamentally changed by the delineation of emotional, social, and ecological intelligences. In this pioneering book, Dan Goleman and his collaborators demonstrate—in vivid and compelling fashion—how education can be transformed through a synthesis of these intelligences."

—**Howard Gardner**, Hobbs Professor of Cognition and Education, Harvard Graduate School of Education

"Gutsy, eloquent, moving, *Ecoliterate* is a masterpiece of motivation and practical guidance. Yes, it is perfect for educators, but it is also perfect for students, parents, grandparents, and all of us yearning to contribute to life on our planet. *Ecoliterate* will help young people discover their own power—and that genie is impossible to get back in the bottle! I love this book."

—**Frances Moore Lappé**, author of *EcoMind: Changing the Way We Think, to Create the World We Want*

"If there is one book on education that needs to be read by every teacher on the planet, *Ecoliterate* is that book. With extremely lucid prose that reflects decades of pioneering work, Goleman, Bennett, and Barlow provide the essential navigation points for a pedagogy that can begin to undo the damage of an antiquated educational and economic system. Theirs is an education for system change. An education infused with hope, passion, and beauty."

—**Laurie Lane-Zucker**, cofounder and former Executive Director, The Orion Society, founder and CEO of Hotfrog

"This book is a treasure. Crucial for educators, a gift to the next generation, a realistic ray of hope for us all. So much ground is covered here—the watersheds where we live as well as the waters of consciousness, how we can change, how we do change, our children finding wisdom while saving this precious earth."

—**Susan Griffin**, author of *Woman and Nature: The Roaring Inside Her*

"Beautifully written, accessible, and urgently important, *Ecoliterate* introduces us to individuals and communities around the country who in very real, practical ways are demonstrating that a new world is possible. No harangue here; this is about hope, embodied in educating our children—head, heart, and hands—to deeply understand how to take care of themselves, their neighbors, and the natural world on which we all depend."

—**Michael Ableman**, farmer and author of *From the Good Earth, On Good Land, and Fields of Plenty*

"In the 21st century, we need a new relationship with nature, a redefinition of both environmentalism and ecoliteracy. In *Ecoliterate*, Daniel Goleman, Lisa Bennett, and Zenobia Barlow offer a roadmap for educators—and the rest of us—to that future, one based on empathy, kinship, natural intelligence, and hope. We're in their debt for showing the way."

—**Richard Louv**, author of *The Nature Principle and Last Child in the Woods*

"Socially and emotionally engaged ecoliteracy can lead to deeply meaningful, inspiring, and effective education. This book shares compelling stories about educators and students who demonstrate the capacity to understand and care about human actions and ecological challenges, and are motivated to take action to make positive differences in today's world. Implementing this vision of education will excite teachers and students, and also improve environments and quality of living in communities across the globe."

—**Roger P. Weissberg**, NoVo Foundation Endowed Chair in Social and Emotional Learning, University of Illinois at Chicago, and president and CEO of the Collaborative for Academic, Social, and Emotional Learning (CASEL)

"*Ecoliterate* is the story of educators with heart, courage, and vision transforming lives, places, and minds. This is 'education' as it should be! Must reading…"

—**David W. Orr**, Paul Sears Professor of Environmental Studies and Politics, Oberlin College

"Preparing our children to live resourcefully in a future few adults can scarcely imagine is daunting for even the most talented educators. Goleman, Bennett, and Barlow convincingly advance a new paradigm for educating children that is founded on the integration of emotional, social, and ecological intelligence. *Ecoliterate* provides a veritable how-to guide for educators to creatively engage young people in the most important ecological issues of the day, helping them gain knowledge of and empathy for all living systems, which is bound to enrich their lives and protect the future of our planet."

—**Gail Connelly**, Executive Director, National Association of Elementary School Principals

"To be ecoliterate is to be equipped to be Earth Citizens, to reach our full potential as human beings. This important book shows us how."

—**Vandana Shiva**, founder, Navdanya International and author of *Soil Not Oil*

"Timely, important, healing, and hopeful—*Ecoliterate* is a 'must read, must implement' guide to a healthy and sustainable present and future."
—**Cheryl Charles**, president and CEO, Children & Nature Network

"The Center for Ecoliteracy has for years been a preeminent thought leader for how we can educate our children in a way that creates generations of earth-stewards. *Ecoliterate* is a much-needed guide for exactly how to accomplish this goal and includes great examples that demonstrate the success of their approach."
—**Oran B. Hesterman, Ph.D.**, president and CEO, Fair Food Network

"*Ecoliterate* is a must-have tool for the twenty-first-century educator seeking ways to connect learning and the environment."
—**Cindy Johanson**, Executive Director, Edutopia

"With our educational system is in drastic need of a makeover, *Ecoliterate* reveals that the key is helping our children understand themselves—their bodies, their communities, and their place in history—in a deep and meaningful way. This book can bring us and our children back to our senses—of taste, touch, and compassion. And its many lessons, resources, and websites help us do that, not someday, but Monday!"
—**Milton Chen**, senior fellow, the George Lucas Educational Foundation and author of *Education Nation*

"Ecological intelligence is more important today than ever before, and educators are the ideal leaders for this movement to attain a greater understanding of both the environment and ourselves. Ecoliterate is a practical and inspirational resource for all educators and communities.
—**Linda Darling-Hammond**, Charles E. Ducommun Professor of Education at Stanford University

ECO

LITERATE

WASHOE COUNTY LIBRARY

3 1235 03751 0521

ECO LITERATE

How **EDUCATORS**
Are Cultivating
EMOTIONAL, SOCIAL,
and **ECOLOGICAL**
INTELLIGENCE

DANIEL GOLEMAN

LISA BENNETT ZENOBIA BARLOW

With Professional Development By
CAROLIE SLY

CENTER FOR ECOLITERACY

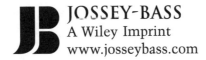

JOSSEY-BASS
A Wiley Imprint
www.josseybass.com

Copyright © 2012 by the Center for Ecoliteracy. All rights reserved.

Published by Jossey-Bass
A Wiley Imprint
One Montgomery Street, Suite 1200, San Francisco, CA 94104-4594—www.josseybass.com

No part of this publication may be reproduced, stored in a retrieval system, or transmitted in any form or by any means, electronic, mechanical, photocopying, recording, scanning, or otherwise, except as permitted under Section 107 or 108 of the 1976 United States Copyright Act, without either the prior written permission of the publisher, or authorization through payment of the appropriate per-copy fee to the Copyright Clearance Center, Inc., 222 Rosewood Drive, Danvers, MA 01923, 978-750-8400, fax 978-646-8600, or on the Web at www.copyright.com. Requests to the publisher for permission should be addressed to the Permissions Department, John Wiley & Sons, Inc., 111 River Street, Hoboken, NJ 07030, 201-748-6011, fax 201-748-6008, or online at www.wiley.com/go/permissions.

Limit of Liability/Disclaimer of Warranty: While the publisher and author have used their best efforts in preparing this book, they make no representations or warranties with respect to the accuracy or completeness of the contents of this book and specifically disclaim any implied warranties of merchantability or fitness for a particular purpose. No warranty may be created or extended by sales representatives or written sales materials. The advice and strategies contained herein may not be suitable for your situation. You should consult with a professional where appropriate. Neither the publisher nor author shall be liable for any loss of profit or any other commercial damages, including but not limited to special, incidental, consequential, or other damages. Readers should be aware that Internet Web sites offered as citations and/or sources for further information may have changed or disappeared between the time this was written and when it is read.

Jossey-Bass books and products are available through most bookstores. To contact Jossey-Bass directly call our Customer Care Department within the U.S. at 800-956-7739, outside the U.S. at 317-572-3986, or fax 317-572-4002.

Wiley publishes in a variety of print and electronic formats and by print-on-demand. Some material included with standard print versions of this book may not be included in e-books or in print-on-demand. If this book refers to media such as a CD or DVD that is not included in the version you purchased, you may download this material at http://booksupport.wiley.com. For more information about Wiley products, visit www.wiley.com.

Library of Congress Cataloging-in-Publication Data
Goleman, Daniel.

  Ecoliterate : how educators are cultivating emotional, social, and ecological intelligence / Daniel Goleman, Lisa Bennett, Zenobia Barlow.
     p. cm.
  Includes bibliographical references and index.
  ISBN 978-1-118-10457-6 (pbk.), ISBN 978-1-118-22397-0 (ebk.), ISBN 978-1-118-23720-5 (ebk.), ISBN 978-1-118-26220-7 (ebk.)
  1. Environmental education--United States. 2. Emotional intelligence--Study and teaching--United States. 3. Social intelligence--Study and teaching--United States. I. Bennett, Lisa, 1959- II. Barlow, Zenobia. III. Title.
  GE70.G66 2012
  363.70071--dc23
                                                                            2012011535

Printed in the United States of America
FIRST EDITION

PB Printing   10 9 8 7 6 5 4 3 2 1

CONTENTS

To Wendy Williams,
for her heartfelt insight and perseverance

ACKNOWLEDGMENTS

The three of us met for the first time in the spring of 2009—and immediately hit it off. We had different backgrounds, experiences, and expertise but also a passionate common ground of interest in the potentially great benefit to be realized in the marriage of social and emotional learning with an ecological view of life. The rich conversations and good humor that we have been privileged to share since then have been both a treasure and a delight. And we have many people to thank for the journey that has ultimately resulted in this book—not the least of which are the many educators from whom we have learned so much. While writing this book, our already abundant store of admiration and respect for the work you do has only grown.

We wish to thank the Center for Ecoliteracy's esteemed board members—Fritjof Capra, physicist and cofounder of the Center for Ecoliteracy, for his brilliant and pioneering work in systems thinking, ecological principles, and leadership; Peter K. Buckley, Center for Ecoliteracy cofounder, for his tremendous insight, skillful blending of hope and realism, and unwavering commitment to education for sustainable living; David W. Orr, who not only originated the phrase "ecological literacy" but has long offered his inimitable inspiration and loyal friendship; and, of course, Wendy Williams, to whom we dedicate this book.

We are tremendously grateful to the entire staff of the Center for Ecoliteracy—especially Carolie Sly, education program director, who contributed the professional development components to the book and generously offered her wise and practical counsel, as well as her good-spirited collegiality, from beginning to end; and Michael K. Stone, senior editor extraordinaire, who contributed a deep knowledge of the subject, a keen ability to spot flaws, and an utterly reliable knack for making things better with the utmost grace.

We also extend our gratitude to Jim Koulias, deputy director, for gently keeping us on schedule and minding the myriad related details with a spirit of joyful support; Alice Lee Tebo, communications coordinator, for her impeccable research, writing, copyediting, and all-around excellence; Jeanne Henry, administrative coordinator, for her energetic and good-natured ability to juggle a multitude of (often simultaneous) requests; and Karen Brown, creative director, for her perceptive reading, engaging, and reflecting on this work. Our thanks also go to Jim Tyler for his exquisite photographs and Leslie Comnes for her dependable research and editing.

We are also immensely grateful to Center for Ecoliteracy friends and funders, especially Marcia Argyris, Erin Eisenberg, Michele Heller, Frances Moore Lappé and Anna Lappé, Dr. Hanmin Liu, Nancy G. Schaub, Brooks Shumway, and Kat Taylor, as well as foundations including The California Endowment, Chilean Forests Preservation Fund, Heller Family Foundation, Orfalea Foundation, Small Planet Fund of RSF Social Finance, TomKat Charitable Trust, Trio Foundation, and S. D. Bechtel, Jr. Foundation.

And of course, to our friends at Jossey-Bass: many thanks to Kate Gagnon and Lesley Iura for believing in this book long before it was one, and for your consistently kind support throughout the process of it becoming one. A special thanks to Tracy Gallagher, Justin Frahm, and Lisa Theobald for such keen and caring eyes, and to Dimi Berkner for sharing with us her extensive experience.

We are also ever grateful to our dear friend Candice Fuhrman for introducing us to the fine people at Jossey-Bass, and to Rowan Foster for all her help keeping we three authors connected.

Needless to say, this book would not have been written without the remarkable people featured in it. We are enormously grateful for the courage and caring with which you have lived your lives and the generosity and openness with which you shared them with us. It was a privilege to visit with each one of you, to learn from your experiences and your reflections, and, in some cases, to receive the further gift of your friendship. To Wendell Berry, Teri Blanton, Cristina Dominguez-Eshelman, Sarah James, Sandy Neumann, Laurette Rogers, Aaron Sharratt, Tony Smith, Allyn Steele, Jane Wholey, Rebecca Wiggins-Reinhard, and Aaron Wolf: our deepest appreciation and admiration.

We also wish to thank the many other individuals who took time to share their experiences and insights with us—including Emily Allen, Rosa Arciniega, Amber Augustine, Melissa Augustine, Angelamia Bachemin, Jo Baker, Kay Brandhurst, Betty Burkes, Mike Coulson, Danny Do, Chris Dorrance, John Elliott, Mallory Falk, Christopher Garcia, Manuel Garcia, Priscilla Garcia, Tom Gardali, Evelyn Gilbert, Sam Gilbert, Dudley Grady Jr., Kent Halla, Rick Handshoe, Xavier

Hernandez, Darwin Jones, Siona LaFrance, Tom Little, Paul Martin, Omar Mateo, Alfredo Matthews, Bev May, Aaron Mihaly, William "Bopper" Minton, Melinda Monterroso, Stephanie Nelson, Dara O'Rourke, John Parodi, Olga Pedroza, Kelly Quane, Christopher Riehle, Matt Roberts, Ricardo Salvador, Eddie Sam, Stanley Sturgill, Patty Tarquino, Shannon Taylor, Paul Vallas, Patty Wallace, Joan Wright-Albertini, and Vanessa Wyant. We offer a special thanks to Jib Ellison and Chris Jordan, who generously gave of their time and powerful perspectives. Thank you to Tom Komer and Alesha Reardon of SouthWings as well for providing an unforgettable flight over what were once the great mountains of eastern Kentucky.

Finally, we offer both thanks and love to our families, for their many shades of support, patience, and good humor.

ECO
LITERATE

INTRODUCTION

FROM BREAKDOWN TO BREAKTHROUGH

Students in a first-grade class at Park Day School in Oakland, California, spent several months transforming their classroom into an ocean habitat, ripe with coral, jellyfish, leopard sharks, octopi, and deep-sea divers (or, at least, paper facsimiles of them). The most in-depth project of their young academic careers, it culminated in one special night when, suited with goggles and homemade air tanks, the boys and girls shared what they learned with their parents. It was such a successful end to their project that several children had to be gently dragged away as bedtime approached.

By the next morning, however, something unexpected had occurred: When the students arrived at their classroom at 8:55 a.m., they found yellow caution tape blocking the entrance. Looking inside, they saw the shades drawn, the lights out, and some kind of black substance covering the birds and otters. Meeting them outside the door, their teacher, Joan Wright-Albertini, explained: "There's been an oil spill."

"Oh, it's just plastic bags," challenged a few kids, who realized that the "oil" was actually stretched-out black lawn bags. But most of the students were transfixed for several long minutes. Then, deciding that they were unsure if it was safe to enter, they went into another classroom, where Wright-Albertini read from a picture book about oil spills.

The children already knew a little bit about oil spills because of the 2010 accident in the Gulf of Mexico—but having one impact "their ocean" made it suddenly personal. They leaned forward, a few with mouths open, listening to every word. When she finished, several students asked how they could clean up *their* habitat. Wright-Albertini, who had anticipated the question, showed them footage of an actual cleanup—and, suddenly, they were propelled into action. Wearing gardening gloves, at one boy's suggestion, they worked to clean up the habitat they had worked so hard to create.

Later, they joined their teacher in a circle to discuss what they learned: why it was important to take care of nature, what they could do to help, and how the experience made them feel. "It broke my heart in two," said one girl. Wright-Albertini felt the same way. "I could have cried," she said later. "But it was so rich a life lesson, so deeply felt." Indeed, through the mock disaster, Wright-Albertini said she saw her students progress from loving the ocean creatures they had created to loving the ocean itself. She also observed them understand a little bit about their connection to nature and gain the knowledge that, even as six- and seven-year-olds, they could make a difference.

It was a tender, and exquisitely planned, teachable moment that reflected what a growing number of educators have begun to identify as a deeply felt imperative: To foster learning that genuinely prepares young people for the ecological challenges presented by this entirely unprecedented time in human history. We are, after all, living at the dawn of an age that has recently been called the "Anthropocene," or "Age of Man." Unlike all the periods that came before, this age is characterized primarily by the ways in which humans are changing nature's systems. And since all life depends on those systems for basic needs, including food, water, and a hospitable climate, there is clearly much at stake. There are also abundant opportunities to practice truly relevant schooling.

This book aims to support and inspire you in your efforts to foster the kind of learning that meets the critical needs of the twenty-first century—and it offers an antidote to the fear, anger, and hopelessness that can result from inaction. It moves, again and again, from breakdown to breakthrough, revealing how the very act of engaging in some of today's great ecological challenges—on whatever scale is possible or appropriate—develops strength, hope, and resiliency in young people. And it presents a model of education for doing so that is founded on a new integration of emotional, social, and ecological intelligence.

"Ecoliterate" is our shorthand for the end goal, while "socially and emotionally engaged ecoliteracy" is the process that we have identified for getting there. We believe the new integration of intelligences it represents offers important benefits both to education and to our societal and ecological well-being. It builds on the successes—from reduced behavioral problems to increased academic achievement—resulting from the movement in education to foster social and emotional learning that has emerged during the past few decades. And it cultivates the knowledge, empathy, and action required for practicing sustainable living.

In the pages that follow, you will see socially and emotionally engaged ecoliteracy in action through stories about innovative educators, artists, activists, scholars, and students who have cultivated these capacities within themselves and are using them effectively to educate others about some of the most critical issues of our time, including food, water, and the two most widely used forms of energy—oil and coal. We recognize that we could have included many other issues, and we could have made many more connections among those we choose. But our goal is simply to illustrate, through these four issues, how socially and emotionally engaged ecoliteracy leads to deeply meaningful, inspiring, and effective education.

Through stories of community leaders putting engaged ecoliteracy into practice, you will meet indigenous Alaskan Sarah James, who is working to protect caribou and native communities from the effects of oil drilling in the Arctic wilderness; Aaron Wolf, a professor of geography who brings a deep spiritual sensibility to his work in helping nations resolve water conflicts; Teri Blanton, a coal miner's daughter who is attracting nationwide attention to the impact of mountaintop mining in Appalachia; and three young leaders, Cristina Dominguez-Eshelman, Rebecca Wiggins-Reinhard, and Aaron Sharratt, who are inspiring people to grow and cook their own food in southern New Mexico.

You will meet teachers and students—from New Orleans, Louisiana; Spartanburg, South Carolina; Oakland, California, and elsewhere—who are demonstrating the capacity to understand and care about the interrelationship between human actions and natural systems, and who are moved to act upon their knowledge, values, and understanding in both small ways and ways as large as saving a mountain. You will learn about five core processes of socially and emotionally engaged ecoliteracy, and some strategies for using this book as a guide for professional development in formal and informal study.

The basic idea is this: At times of instability in a system—be it a school, a nation, or the biosphere—there is always the possibility of breakthrough to new forms and ways of thinking and acting.[1] In these times of instability—in our schools, our nation, *and* our biosphere—this book reflects our core belief that educators are ideally situated to lead a breakthrough to a new and enlivening ecological sensibility for the twenty-first century.

WHY WE NEED A NEW ECOLOGICAL SENSIBILITY

We humans, of course, have always affected the natural world on which we depend. But with 7 billion of us—up from about 1.6 billion in 1900—now tapping the Earth's resources, we are having an impact like never before. Consider the growing scarcity of fresh drinking water, the decline of healthy soil in which to grow our food, and global climate change. With the world's population projected to increase to 9 billion by the middle of this century, we have to ask, Is there a breaking point?[2]

In 2009, Johan Rockström, director of the Stockholm Environment Institute in Sweden, set out to answer that question with the help of more than two dozen of the world's leading scientists, including Nobel Laureate Paul Crutzen and NASA climate scientist James Hansen. They identified nine life-support systems essential for human survival, including biochemical cycles such as carbon and water, and physical circulation systems such as our climate and oceans.[3]

For each of these life-support systems—later dubbed "Earth's Nine Lives"[4]—the scientists described a safe zone within which human development can securely operate. Somewhere beyond these boundaries—but no one knows exactly how far—we risk triggering "abrupt or irreversible environmental changes that would be deleterious or even catastrophic for human well-being."

So how are we doing? For seven of Earth's nine "lives," the evidence is clear: We have far exceeded the safe boundary levels of two of these life-support systems (biodiversity and the nitrogen cycle).[5] We recently surpassed the boundary on a third (climate change). And we are projected to hit the boundary on three more (ocean acidification, freshwater cycles, and land use) by mid-century. (Rockström and his colleagues were unable to determine boundary levels for chemical pollution and aerosol loading.)

Thankfully, none of this means that the sky is falling—yet. On the condition that we have not transgressed these boundaries for too long, the authors suggest, humanity appears to have some room to maneuver. But there is considerable uncertainty as to how long or far we can go beyond the boundaries and still be able to return to safe levels. Consequently, we need to urgently put the brakes on before we reach the tipping points where systems spin out of control and crash.

The good news is there is reason to be hopeful. After all, humans have shown that when they grasp that their actions are threatening one of life's support systems—and do so on a deep level that taps both the cognitive and the feeling ways of knowing—they can take effective action. (The response to the hole in the ozone layer is a case-in-point. "The ozone hole that formed in the stratosphere over Antarctica in the 1970s was a classic example of an environmental tipping point," Fred Pearce wrote in *New Scientist* magazine. Nobody had seen it coming. But once people realized the severity of the problem, they acted quickly to ban most ozone-destroying chemicals, and we began moving back into the safe zone on this essential life-support system.) Developing emotional, social, and ecological intelligence can help us now effectively address the remaining threats to our Earth's life support systems.

MAKING THE CONNECTIONS

However important ecological sensibility is today, the fact is that most of us do not truly grasp how our everyday actions—our engagement in the systems of energy, agriculture, industry, commerce, and transportation on which we rely—can threaten the health and well-being of the Earth. For example, ask your students (maybe even your colleagues): Where does your electricity come from? What is the connection between your container of apple juice and the lives of baby sea birds thousands of miles away? How many environmental impacts result from the production of the steel used to make your "eco-friendly" water bottle? (Hint: It is a four-digit answer.[6])

Very few people know the answers to these or thousands of similar questions for one fairly straightforward reason: The complexity of the web of connections that characterize our global society has created a vast collective blind spot about the effects of human behavior on natural systems. Imagine, for a moment, what it would be like if you, your students, and their families lived in a small, isolated community—say, in the Arctic, the Sahara Desert, or the mountains of Tibet—where you relied on each other alone for all your basic needs. Say food, for example, was not flown in from half a world away but grown and shared right where you live. If your community decided to farm in ways that were expedient one year but failed to leave the soil healthy for the next, experience would soon teach you about the interconnections between human behavior and the health of natural systems. And you would be much more likely to be aware that the resilience of future generations ultimately depends on the wise use of natural resources and adaptation to our ecological niche.

Today, however, the vast majority of us do not live in small, isolated communities but in cities and suburbs,[7] where we depend on people and processes from around the world to meet most of our basic (and not so basic) needs. Our use of resources and the ensuing ecological impacts are dispersed across the entire planet—often seeming invisible or too far away for us to fully recognize.

Moreover, even when a young person's knowledge and empathy have been awakened, it can be a magnificent challenge to help him or her understand how to make a positive difference in the world today. Yet it is a magnificent challenge that we believe can be met through the cultivation of socially and emotionally engaged ecoliteracy.

FROM EMOTIONAL TO ECOLOGICAL INTELLIGENCE

Nearly thirty years ago, Howard Gardner's groundbreaking book, *Frames of Mind: The Theory of Multiple Intelligences*, effectively moved a generation of educators beyond the narrow notion of "IQ" that had dominated much of the twentieth century. Schools, he argued, must not educate to one narrow notion of intelligence but to seven (later eight) forms of intelligence: bodily-kinesthetic, interpersonal, verbal-linguistic, logical-mathematical, intrapersonal, visual-spatial, musical, and, most recently, naturalistic.

In 1995, Daniel Goleman explored another significant dimension of intelligence in his book *Emotional Intelligence: Why It Can Matter More than IQ*. Drawing on brain and behavioral research, he examined the factors at work when people of high IQ flounder and those of modest IQ do surprisingly well. Those factors included five critical aspects of emotional intelligence that could be nurtured in schools: the abilities to know one's emotions, manage those emotions, motivate oneself, recognize emotions in others, and develop successful relationships.

In his next book, *Social Intelligence: The Revolutionary New Science of Human Relationships*, Goleman advanced a second model of intelligence that comes into play in our relationships with others. He reported on research demonstrating that our brains make us "wired to connect," and showed how this, too, is a key ingredient for success in life—and a "neural key to learning."

These two books helped inspire the rapid growth of a movement for social and emotional learning (SEL), which emphasizes the development of knowledge, attitudes, and skills around these intelligences. To date, tens of thousands of schools have adopted social and emotional learning programs. The state of Illinois currently has comprehensive, free-standing learning goals and benchmarks for social and emotional learning. And four other states—Pennsylvania, New York, Washington, and Kansas—are considering or adopting similar policies. One reason that educators have been drawn to SEL is that research has shown that nurturing these understandings, values, and abilities contributes to significant improvements in academic achievement. (See "Benefitting Academic Achievement and Ecological Well Being" below.)

Building on this work, Goleman introduced a third and related kind of intelligence in the 2009 book, *Ecological Intelligence: The Hidden Impacts of What We Buy*. While social and emotional intelligence extend students' abilities to see from another's perspective, empathize, and show concern, ecological intelligence applies these capacities to an understanding of natural systems and melds cognitive skills with empathy for all of life.

The findings of these studies of emotional, social, and ecological intelligences paralleled the practical experience of the Center for Ecoliteracy. For the past twenty years, the Center for Ecoliteracy has applied theories of leadership and systems thinking to its mission of advancing education for sustainable living in primary and secondary schools, through a pedagogy steeped in the process of social and emotional learning. Cofounded by systems thinker Fritjof Capra, executive director Zenobia Barlow, and environmental philanthropist Peter Buckley, the Center has launched pioneering initiatives involving creation of school gardens, school lunch reform, habitat restoration and preservation, and integration of sustainability into teaching and learning.[8] Throughout its work with educators in thousands of communities across the globe, the Center has recognized emotional, social, and ecological intelligence as essential perspectives that develop empathy, mindfulness, and new modes of cooperation to help communities live sustainably.

PUTTING THE PIECES TOGETHER

The model of education we present in this book takes the cultivation of emotional and social intelligence as its foundation and expands this foundation to integrate ecological intelligence. But rather than conceive of these as three separate types of intelligences, we posit emotional, social, and ecological intelligence as essential dimensions of our universal human intelligence that simply expand outward in their focus: from self, to others, to all living systems. We also conceive of these intelligences in a dynamic relationship with each other: Cultivate one, and you help cultivate the others.

The practice of socially and emotionally engaged ecoliteracy, as the stories in this book illustrate, can take many forms. Nevertheless, we identify two core dimensions as through-lines. One is affective, or related to emotions: namely, empathy for all forms of life. By "empathy," the ability to understand and share the feelings of another, we do not intend to imply that plants, for instance, have feelings. Rather, our intent is to encourage a sense of caring that is not restricted to other human beings but extends to all forms of life.

The other through-line is cognitive, or related to how we think: that is, understanding how nature sustains life. Since life began, Earth's ecosystems have developed ways of supporting the great web of human and nonhuman life through certain patterns and processes, such as cycles, networks, and nested systems—all of which reflect the fundamental fact, as Center for Ecoliteracy cofounder Fritjof Capra put it, that "nature sustains life by creating and nurturing communities." To understand how nature sustains life, then, requires the capacity for systems thinking, or the ability to perceive how the different aspects of a living system exist, both in relationship to one another and relative to the whole that is greater than its parts.

But how, it would be reasonable to ask, can anyone truly develop the capacity to understand all the ways in which human systems interact with natural systems and act upon that knowledge? The answer is simple: We can't. Not alone. Rather than residing within a single individual, ecological intelligence is inherently *collective*. Socially and emotionally engaged ecoliteracy, therefore, encourages us to gather and share information collectively, and to collectively take action to foster sustainable living. This makes school communities—which, like ecosystems, come to life through networks of relationships—ideal places to nurture this new and essential ecological sensibility.

BEYOND OUR COLLECTIVE BLIND SPOT

The past decade has seen an explosion of new information and resources that teachers can use to help students visualize the interrelationships between human actions and living systems. These tools include the Ecological Footprint (www. myfootprint.org), Fooducate (www.fooducate.com), and Google Earth (www. google.com/earth/). New Life Cycle Assessment (LCA) tools can help you operationalize systems thinking by revealing all of the many points of interaction and processes involved in, for example, the production and consumption patterns of something as prevalent as a cell phone.

One example of an LCA tool is GoodGuide (www.goodguide.com), developed by University of California, Berkeley, Professor Dara O'Rourke. GoodGuide evaluates most mainstream consumer products available in the United States, based on their impacts on health, the environment, and society (including such factors as employee working conditions). The assessment is a complex process, involving teams of scientists sometimes analyzing up to hundreds of factors; but the tool simplifies the process by serving up a simple rating for each item analyzed.

In the chapters that follow, we will present tools and strategies that illuminate what lies beyond our collective blind spot about the interrelationships between human actions and living systems, presenting educators with opportunities to teach about these connections in ways that are vivid, concrete, and deeply relevant to today's young people.

BENEFITTING ACADEMIC ACHIEVEMENT AND ECOLOGICAL WELL-BEING

Social and emotional learning was embraced by many schools on the promise that helping children develop the capacities for self awareness, self management, social awareness, and relationship management would increase their likelihood of success in school and in life. Now, extensive research shows that these programs do lead to important student gains and reduced risks for failure. For example, a 2011 meta-analysis by the Collaborative for Academic, Social, and Emotional Learning of 213 social and emotional learning programs concluded that these programs improve students' achievement test scores by 11 percentile points—and lead to even greater gains in improved attitudes and positive classroom behaviors,[9] as well as reductions in conduct problems and disciplinary actions.[10]

Research studies examining the influence of an environment-based context have revealed similar encouraging findings with respect to academic performance.[11] In the two decades that the Center for Ecoliteracy has worked with schools to support education for sustainable living, it has found that socially and emotionally engaged ecoliteracy advances both teacher and student involvement and achievement through hands-on, experiential, contextual learning in the natural world and community.

With this as our foundation, we are confident that school communities are the ideal places—and educators the ideal leaders—to guide a breakthrough to a new, enlivening, and much needed ecological sensibility. In the pages that follow, we will show how you can cultivate teaching and learning that help young people develop the capacity to perceive, understand, and care about the interrelationship between the natural world and human actions—and then apply that understanding to guide individual and collective human action toward the wiser use of natural resources and adaptation to our true ecological niche, which is, of course, nothing less than our interconnected world. As author and farmer Wendell Berry has said, "Ultimately, what we have to be talking about is not what we are against but what we are for, and that is life not lived against the possibility of life."[12]

5 Practices of Emotionally and Socially Engaged Ecoliteracy

With a goal of nurturing students to become ecoliterate, the Center for Ecoliteracy has identified five vital practices that integrate emotional, social, and ecological intelligence. We work to inspire teachers to use a variety of learning opportunities that help students consider and apply these practices in a diverse range of contexts. These practices allow students to strengthen and extend their capacity to live sustainably.

1 Developing Empathy for All Forms of Life encourages students to expand their sense of compassion to other forms of life. By shifting from our society's dominant mindset (which considers humans to be separate from and superior to the rest of life on Earth) to a view that recognizes humans as being members of the web of life, students broaden their care and concern to include a more inclusive network of relationships.

2 Embracing Sustainability as a Community Practice emerges from knowing that organisms do not exist in isolation. The quality of the web of relationships within any living community determines its collective ability to survive and thrive. By learning about the wondrous ways that plants, animals, and other living things are interdependent, students are inspired to consider the role of interconnectedness within their communities and see the value in strengthening those relationships by thinking and acting cooperatively.

③ Making the Invisible Visible assists students in recognizing the myriad effects of human behavior on other people and the environment. The impacts of human behavior have expanded exponentially in time, space, and magnitude, making the results difficult if not impossible to understand fully. Using tools to help make the invisible visible reveals the far-reaching implications of human behavior and enables us to act in more life-affirming ways.

④ Anticipating Unintended Consequences is a twofold challenge of predicting the potential implications of our behaviors as best we can, while at the same time accepting that we cannot foresee all possible cause-and-effect associations. Assuming that the ultimate goal is to improve the quality of life, students can adopt systems thinking and the "precautionary principle" as guidelines for cultivating a way of living that defends rather than destroys the web of life. Second, we build resiliency by supporting the capacity of natural and social communities to rebound from unintended consequences.

⑤ Understanding How Nature Sustains Life is imperative for students to cultivate a society that takes into account future generations and other forms of life. Nature has successfully supported life on Earth for billions of years. Therefore, by examining the Earth's processes, we learn strategies that are applicable to designing human endeavors.

Five Practices of Emotionally and Socially Engaged Ecoliteracy

Educators who cultivate the following five practices of socially and emotionally engaged ecoliteracy—for both themselves and their students—have the profound capacity to help create and sustain healthier relationships with other people and the planet. These are, of course, not the only ways to do so. But we believe these practices, described below and illustrated in the stories to follow, offer a strong foundation for becoming ecoliterate. Each can be nurtured in age-appropriate ways for students, ranging from pre-kindergarten through adulthood, and help promote the cognitive and affective abilities central to the integration of emotional, social, and ecological intelligence.

DEVELOPING EMPATHY FOR ALL FORMS OF LIFE

People who are ecoliterate cultivate compassion toward other forms of life. This ability to feel empathy often stems from a deep understanding that humans are part of a broader community that includes all living beings.

At a basic level, all organisms—including humans—need food water, space, and conditions that support dynamic equilibrium to survive. By recognizing the common needs we share with all organisms, we can begin to shift our perspective about who we are relative to other species and life forms. We can move from a view of humans as separate and superior to a more authentic view of humans as members of the natural world. From that perspective, we can expand our empathic concern to consider the quality of life of other life forms, feel genuine concern about their well-being, and act on that concern.

Most young children exhibit care and compassion toward other living beings. This is one of several indicators suggesting that human brains are wired to feel empathy and concern for other living things. Teachers can nurture this capacity to care by creating class lessons that emphasize the important roles that plants and animals play in sustaining the web of life. Empathy also can be developed through direct contact with other living things. Teachers often nurture caring among their students by keeping live plants and animals in the classroom; taking field trips to nature areas, zoos, botanical gardens, and animal rescue centers; and involving students in field projects such as habitat restoration. In Chapter Six of this book, you will read an inspiring example of fourth graders who chose an unlikely endangered species, fell in love with it, and participated in restoring its habitat.

Another way teachers can help develop empathy for other forms of life is by studying indigenous cultures. Most of these cultures share a profound reverence for all living creatures, especially for those that co-exist with them in and around where they live. From early Australian Aboriginal culture to the Gwich'in First Nation in the Arctic Circle, traditional societies have viewed themselves as intimately connected to plants, animals, the land, and the cycles of life. This worldview of interdependence guides daily living, contributing to these societies' ability to survive, frequently in delicate ecosystems, for thousands of years. Chapter Three, "The Heart of the Caribou," offers a moving example of the relationship between an indigenous community and the living beings with which they share their homeland. The study of indigenous cultures is found throughout academic standards from kindergarten through twelfth grade. By focusing on their relationship with their surroundings, students learn how a society lives when it values other forms of life.

EMBRACING SUSTAINABILITY AS A COMMUNITY PRACTICE

As you read this book, you will notice that the remarkable people portrayed here practice ecoliteracy within the context of their community. They are well aware that emotional and social intelligence flourish through their interconnectedness as they join to focus on issues about which they are passionate.

Communities represent a core pattern of organization for surviving over time. When we turn to nature, we find that healthy communities of living organisms are diverse, have a strong network of relationships, and are resilient. Life in nature does not survive in isolation.

When we consider human communities, we can point to numerous examples of those that strive to be sustainable. Many are ancient indigenous societies, and some are more recent intentional communities located throughout the world. These kinds of communities, which are diverse in their structure and function, frequently offer inspiring models of sustainable living.

One commonly held value among human communities that practice sustainable living is a high regard for the common good. Recognizing that "we are all in this together," sustainable communities endeavor to create general conditions that are to the advantage of both people and other life. In daily living, this translates to preserving soils, habitats, and water for the long term; practicing energy conservation and prioritizing renewable resources; and keeping waste to a minimum. It also includes creating economic and governmental systems that can sustain the community if an unanticipated disruption occurs, by decentralizing essential goods and services and building in redundancy so if one part of the system fails, other parts are able to keep operating.

Teachers and students participate in a variety of communities at home, at school, and within the wider context of their lives. Many schools strive to build community on campus and sometimes with the broader local populace around them.

The notion of sustainability as a community practice, however, embodies some characteristics that fall outside most schools' definitions of themselves as a "community." For example, by examining how their community provisions itself—from school food to energy use—students can contemplate whether their everyday practices reflect valuing the common good. In Chapter Four, "Beyond Whining," you will read about high school students in New Orleans who are determined to change the school system so that it models sustainable living.

Other students might follow the approach taken in New Orleans and gather data about the sources of their energy and the amount they use and then survey their peers by asking, "How might we change the way we use energy so that we are more resilient and reduce the negative impacts on people, other living beings, and the planet?" As the New Orleans story shows, these projects can give students the opportunity to begin to build a community that values diverse perspectives, the common good, a strong network of relationships, and resiliency.

MAKING THE INVISIBLE VISIBLE

Historically—and for some cultures still in existence today—the path between a decision and its consequences was short and visible. If a homesteading family cleared their land of trees, for example, they might soon experience flooding, soil erosion, a lack of shade, and a huge decrease in biodiversity. These days, we often don't see the far-reaching implications of many of our actions—they seem to be invisible. For example, clearing rainforests in South America for planting cash crops such as coffee impacts people and the natural world in numerous ways, as carbon is released into the atmosphere, contributing to the greenhouse effect. Local communities are often displaced, biodiversity is reduced, the soil becomes eroded and depleted of nutrients, and nearby water is polluted due to increased runoff. Yet, as we sip our morning coffee in the United States, the consequences of growing coffee beans in formerly rainforest-covered areas of South America are largely invisible to us. It is too far away for us to fathom.

People who practice engaged ecoliteracy realize that our global economy has created blinders that shield us from experiencing the far-reaching implications of our actions. As we have increased our use of fossil fuels, for instance, it has been difficult (and remains difficult for many people) to believe that we are disrupting something on the magnitude of the Earth's climate. Although some places on the planet are beginning to see evidence of climate change, most of us experience no changes. We may notice unusual weather, but daily weather is not the same as climate disruption over time. In another example, as we develop and use new chemicals, it is unfathomable for most of us to envision the implications for future generations. If we endeavor to develop ways of living that are more life-affirming, we must find ways to make the things that seem invisible visible.

Educators can help make the invisible visible to students by using a number of strategies. As mentioned, phenomenal web-based tools can be used to allow students to visualize that which they cannot see. Google Earth, for example, allows students to "travel" virtually and view the landscape in other regions and countries. Technological applications such as GoodGuide and Fooducate are able to cull from a great deal of research and "package" it in easy-to-understand formats that inform us about the impact of certain household products on our health, the environment, and social justice. Through social networking websites, students can also communicate directly with citizens of distant areas and learn firsthand what the others are experiencing that is invisible to most students. Finally, some students, such as those discussed in Chapter Two, "Taking a Power Trip," can directly observe places that have been quietly devastated as part of the system that provides most of us with energy.

ANTICIPATING UNINTENDED CONSEQUENCES

This practice is twofold: It involves better ways of predicting the potential implications of our behavior while simultaneously creating strategies that compensate for the fact that we cannot foresee all the possible effects of our actions.

Many of the environmental crises that we face today are the unintended consequences of human behavior. For example, we have experienced many unintended but grave consequences of developing the technological ability to access, produce, and use fossil fuels. These new technological capacities have been largely viewed as progress for our society. Only recently has the public become aware of the downsides of our dependency on fossil fuels, such as pollution, suburban sprawl, international conflicts, and climate change.

As a teacher, you can use a couple of noteworthy strategies for teaching students to anticipate unintended consequences. One strategy—the precautionary principle—can help students think about possible unanticipated consequences of everyday human behavior and then shift their thinking and behavior. Several formulations of the precautionary principle exist, but their collective message is the same: When an activity threatens to have a damaging impact on the environment or human health, precautionary actions should be taken regardless of whether a cause-and-effect relationship has been scientifically confirmed. Historically, the people concerned about possible negative impacts of new products, technologies, or practices were expected to prove scientifically that harm would result in order to impose restrictions on their use. Rather than giving the benefit of the doubt to the producers, the precautionary principle (which is now in effect in many countries and in some places in the United States) places the burden of proof on the producers to demonstrate harmlessness and accept responsibility should harm occur.

Throughout this book, you will see unanticipated consequences of *not* following the precautionary principle. Any one of these stories could serve as a lesson for older students to identify what the outcomes might have been if decision-makers had heeded the precautionary principle and refrained from acting until possible consequences were better understood. After practicing applying the precautionary principle, in retrospect, students can apply it to some of the current decisions facing society today, such as the use of genetically modified crops or electric vehicles or wind farms.

Five Practices of Emotionally and Socially Engaged Ecoliteracy   **15**

Another strategy for anticipating unintended consequences is to shift from analyzing a problem by reducing it to its isolated components, to adopting a systems thinking perspective that examines the connections and relationships among the various components of the problem. These two approaches to thinking are not either/or choices in how we approach an issue; they represent a continuum of ways to take in and make sense of information. However, modern Western society has long valued reductionist, linear thinking over systems thinking. Consequently, we have often undermined our collective ability to anticipate the unintended consequences of our actions by focusing on the parts at the expense of the whole. Students who can apply systems thinking are usually better at predicting possible consequences of a seemingly small change to one part of the system that can potentially affect the entire system. One easy method for looking at a problem systemically is by mapping it and all of its components and interconnections. It is then easier to grasp the complexity of our decisions and foresee possible implications.

Finally, no matter how adept we are at applying the precautionary principle and systems thinking, we will still encounter unanticipated consequences of our behaviors. Building resiliency (for example, by moving away from mono-crop agriculture or by creating local, less centralized food systems or energy networks) is another important strategy for survival in these circumstances. We can turn to nature and find that the capacity of natural communities to rebound from unintended consequences is vital to survival.

UNDERSTANDING HOW NATURE SUSTAINS LIFE

You can recognize emotionally and socially engaged ecoliterate people by how they participate in their everyday lives. Their ways of living reflect their understanding of how nature sustains life. Recognizing that nature has sustained life for eons, they have turned to nature as their teacher and learned several crucial tenets. Three of those tenets are particularly imperative to ecoliterate living.

First of all, ecoliterate people recognize that they are members of a web of diverse relationships within their communities and beyond. They have learned from nature that all living organisms are members of a complex, interconnected web of life and that those members inhabiting a particular place depend upon their interconnectedness for survival. Teachers can foster an understanding of the diverse web of relationships within a location by having students study that location as a system.

Second, ecoliterate people tend to be more aware that systems exist on various levels of scale. In nature, organisms are members of systems nested within other systems, from the micro-level to the macro-level. Each level supports the others to sustain life. When students begin to understand the intricate interplay of relationships that sustain an ecosystem, they can better appreciate the implications for survival that even a small disturbance may have, or the importance of strengthening relationships that help a system respond to disturbances. Students can apply their understanding of nested systems to relationships within their school, community, and other ecosystems.

As you will read in Chapter Seven, "Changing a Food System, One Seed at a Time," when savvy and motivated young people understand the nested systems in their community, they can influence their community to become more resilient.

Finally, ecoliterate people collectively practice a way of life that fulfills the needs of the present generation while simultaneously supporting nature's inherent ability to sustain life into the future. They have learned from nature that members of a healthy ecosystem do not abuse the resources they need in order to survive. They have also learned from nature to take only what they need and to adjust their behavior in times of boom or bust. This requires that students learn to take a long view when making decisions about how to live.

These five core practices inform the approach to pedagogy described in Section Two of this book. They integrate emotional, social, and ecological intelligence and, when practiced together, manifest a whole that is much more than the sum of its parts.

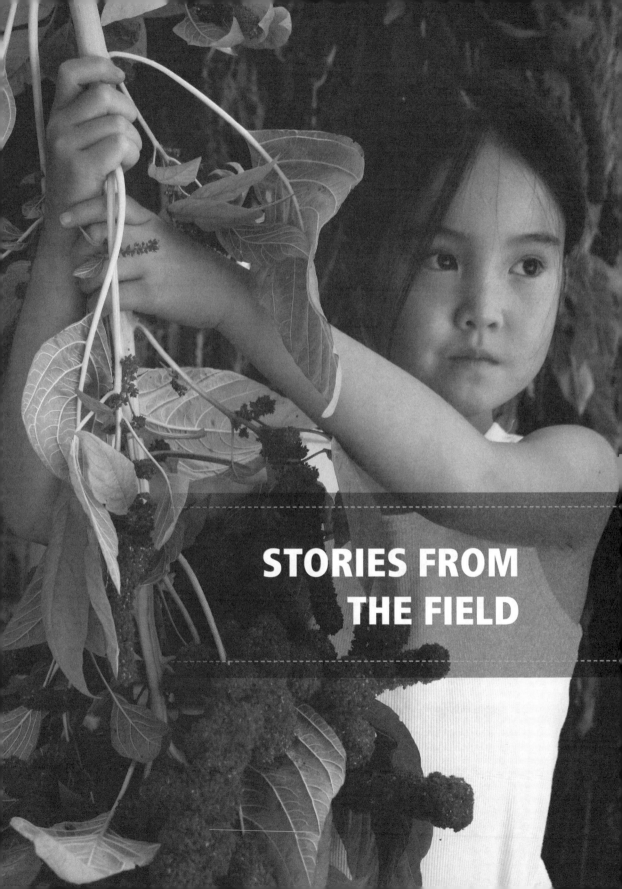

**STORIES FROM
THE FIELD**

PART ONE

STANDING STRONG
ON A COAL MOUNTAIN

Lessons from a Coal Miner's Daughter

Teri Blanton and Wendell Berry, Kentucky

This story reveals how a group of everyday citizens—committed to protecting the health of their community and the Appalachian ecosystem—find strength, influence, and friendship by embracing sustainability as a community practice: one of the five practices of socially and emotionally engaged ecoliteracy. As you read, notice the diverse backgrounds of the individuals and the ways in which mountaintop removal mining has affected their lives.

A T 9:45 ON A COLD FEBRUARY MORNING in 2011, a dozen Kentucky residents—including educators, writers, retired coal miners, and one rather determined coal miner's daughter—gathered in the parking garage of the state capitol. Each carried a red lunch bag containing sandwiches, chips, and apples, along with their driver's licenses in their pockets for identification. Some carried supplies of prescription medications. The best known among them, acclaimed seventy-six-year-old author and farmer Wendell Berry, had a toothbrush tucked in the inside pocket of his suit jacket. After decades of opposition to mountaintop mining in Appalachia, one of the most biologically important regions in the United States,[1] the group had arrived at the capitol to confront the governor. And because they considered mountaintop removal mining both a human rights and an environmental issue, they were prepared to be arrested.

As an unidentified man left the building, fifty-four-year-old Teri Blanton ran ahead to catch the door. Then the group walked in, through a pedestrian tunnel, and up the stairs, pausing a few yards from Governor Steve Beshear's office.

"Ready?" Blanton asked.

"Ready," they answered.

Inside the governor's office, someone informed the receptionist that they had sent a letter announcing their intention to visit but had received no response. The receptionist went behind a closed door and then returned to say that the governor was busy, but she would see if someone else could meet with them.

"Will that be all right?" she asked.

"I think we'll just wait here to meet with the governor," Blanton replied, as others settled in on couches and chairs.

A few minutes later, the receptionist asked if they would move down the hall to a conference room.

"I think we'll just stay here and wait for the governor," repeated Blanton.

Ten minutes later, Chief of Staff Mike Haydon offered to meet with them.

"I think we'll just wait for the governor," Blanton said for a third time.

Blanton, who received a 2010 Rainforest Action Network award for outstanding leadership to protect the environment, lost a brother to a coal mining accident and friends to cancer that she believes was coal-related. In her hometown of Dayhoit, Kentucky, she watched her children tromp through coal muck every morning to catch the school bus.

But, she insists, "I ain't nobody's damn victim." Indeed, Blanton has stood up to coal companies, addressed hundreds of people at rallies, challenged a United States senator on MSNBC, and helped inspire an investigation that led to the designation of an EPA Superfund site.

Called "the Erin Brockovich of mountaintop mining," Blanton has educated people about this issue on the local, state, and national level, displaying a home-grown brand of emotionally and socially engaged ecological intelligence.[2] She especially seeks to help those of us who do not live anywhere near Kentucky or other coal mining states to recognize our personal connection to the Appalachian region. She reminds us that every time we flip on the lights in our home, office, or school, the electricity we use is often generated from burning coal.

THE ROLE OF COAL TODAY

The United States is the world's largest producer of coal, the most significant source of electricity today. It fuels the generation of 45 percent of the electricity used in this country and 40 percent of the electricity used worldwide, with global coal consumption projected to increase 53 percent by 2035.[3]

But if coal mining conjures up images of men going down into underground mines, to blast and dig out the sedimentary rock, that's not the way it typically happens today. Sixty percent of coal mined in the United States is now acquired through surface mining, which uses a variety of techniques to remove soil, rock, and entire ecosystems to access the minerals below the surface. This includes strip mining, open pit mining, and mountaintop mining (or mountaintop removal)—the most controversial and destructive form of all.[4] Significantly, for the coal industry, mountaintop mining is also the most profitable method because it relies more on machinery and explosives than people.

HOW TO MINE A MOUNTAIN

Mountaintop mining—the most profitable albeit most destructive form of coal mining—is carried out in six basic steps:

1. Use bulldozers to clear trees and level the mountaintop.

2. Drill small holes through the dirt and rock and drop in powerful explosives that blow as much as 800 feet off the mountaintop to reveal the coal seams below the surface.[5]

3. Excavate the coal using power shovels or a 2,000-ton, twenty-story-high dragline, one of the largest machines in the world.[6]

4. Bulldoze the remains of the shattered mountains and their ecosystems (what the industry calls "overburden") into the valleys and streams below.

5. Gather the coal and transport it to a plant to be washed before shipping. Then dump the liquid waste, or slurry—containing arsenic, lead, mercury, magnesium, and selenium— into a hillside dam.

6. "Reclaim" the area. By law, the coal company is required to create "useful landscapes" after a mining operation is completed. This can include, for example, replacing soil, replanting trees, and restoring the basic contours of the landscape that existed before the mountains were blasted and bulldozed away. Put another way, they are obliged to remake nature's design—a feat few find plausible.

Mountaintop mining is largely confined to the Appalachian Mountains, one of the economically poorest—and ecologically richest—regions in the United States. (If mountaintop mining occurred in more affluent parts of the country, Robert F. Kennedy Jr., has said, it would likely lead to jail time.[7]) The oldest mountain range in North America, the Appalachian Mountains are home to an extraordinary diversity of flora and fauna, thanks to the legacy of the Ice Age and a generally mild climate.[8]

As a result of mountaintop mining, an estimated 500 mountaintops, one million acres of forest, and 2,000 miles of streams have been destroyed since the mid-1980s.[9] In 2010, a team of researchers also found higher rates of hospitalizations and deaths due to heart, lung, and kidney problems in mountaintop mining regions than elsewhere in the United States.[10] And a number of mining communities have become ghost towns, as residents have sold their homes to coal companies rather than live amid the noise, pollution, and general devastation of the landscape. In Lindytown, West Virginia, for example, only two occupied houses remain in what was once a small mining town: one belonging to an elderly woman suffering from Alzheimer's, whose family felt it would be too disorienting to move, and another belonging to her son.[11]

The burning of coal—acquired through mountaintop mining or, for that matter, any other means—also threatens a number of the life-support systems on which we depend. For example, the United States alone produces close to two billion tons of carbon dioxide (CO_2) per year from coal-burning power plants.[12] And CO_2 is a significant contributor to climate change, the increasing acidity of the oceans, and interference with Earth's nitrogen cycle, leading to "dead zones" in oceans and rivers. (Dead zones are areas where oxygen levels are too low to support marine life. The Gulf of Mexico, for example, is home to a dead zone as large as the state of New Jersey.)

"NOBODY'S DAMN VICTIM"

Blanton, Berry, and a growing number of others have been expressing their opposition to mountaintop mining for years. Berry, in fact, says that he has voiced his opposition to all surface mining for nearly a half century. Reflecting on the destruction inflicted on his home state during that time, Berry wrote, "This is a history by any measure deplorable, and a commentary sufficiently devastating upon the intelligence of our politics and our system of education."[13] By way of example, he pointed to a lack of understanding about the difference between the long-term value of a well-maintained forest ecosystem and the short-term gain of coal mining—likening the difference to "using a milk cow, and her daughter and granddaughters after her for a daily supply of milk, renewable every year—or killing her for one year's supply of beef."[14]

DECONSTRUCTING THE MEANING OF "CHEAP"

Coal is the most common source of electricity in the world because it is plentiful—and one of the cheapest forms of energy, per kilowatt-hour (kWh). But there is a great difference between *price* and *cost*, as a recent life cycle analysis of coal revealed.

"Each stage in the life cycle of coal—extraction, transport, processing, and combustion—generates a waste stream and carries multiple hazards for health and the environment," wrote Paul R. Epstein, the late associate director of Harvard Medical School's Center for Health and the Global Environment.[18]

These costs, known as hidden or external costs, are not paid for by coal manufacturers and therefore are not reflected in the price. They are instead transferred to society at large.

"We estimate that the life cycle effects of coal and the waste stream generated are costing the U.S. public a third to over one-half of a trillion dollars annually," Epstein wrote in the February 2011 issue of *Annals of the New York Academy of Sciences*. Moreover, many of these hidden costs are cumulative.

His report identified health, economic, and environmental hazards at various stages of coal's life cycle, from extraction and processing to transportation and combustion. Among the economic

Moreover, he has said, our education system plays a role in the perpetuation of ecological destruction, because it is based on the faulty premise of an economy that externalizes health, environmental, and other costs: "The change that is called for is a shift from the economy to the ecosphere as the basis of curriculum, teaching, and learning." The ecosphere, Berry says, is the true basis and context for any economy.[15]

After decades of opposition, Berry announced in 2008 that he was losing patience, and that it was time to invoke civil disobedience. "If your government will not rise to the level of common decency, if it will not deal fairly, if it will not protect the land and people, if it will not fully and openly debate the issues, then you have to get in the government's way," he declared at a rally against mountaintop mining that year.[16]

The final straw came in 2010. After the Environmental Protection Agency (EPA) announced that it would block eleven mining permits out of concern for the impact of mining on the state's waterways, Governor Beshear partnered with the coal industry to sue the agency.[17] This action, as Blanton put it, made clear that the government put coal interests above the interests of people, land, and water—and it was time to get in the government's way.

With this as the motivation for their visit to the state capitol in February 2011, Blanton, Berry, and the others were determined to wait for the governor to listen to them, no matter how long it took. As they waited, several reporters arrived and asked, "Do you really think talking to a governor who has sided with the coal industry will work?"

impacts are the cost of federal and state subsidies of the coal industry, decline of property values as a result of mountaintop mining, damage to farmland and crops resulting from pollution, loss of tourism, and costs to taxpayers of environmental monitoring and mitigation.

Among the health impacts are increased rates of mortality and morbidity resulting from elevated levels of soot and other fine particles in the air, and higher rates of sudden infant death syndrome.

Among the environmental impacts are methane emissions that lead to climate change, loss and contamination of streams, destruction of local habitats and biodiversity, acid mine drainage, air pollution, heavy metal pollution, ozone emissions, soil contamination from acid rain, and destruction of marine life from mercury pollution and acid rain.

In short, if the price of coal reflected its true cost, it would nearly triple—making coal no longer one of the cheapest, but one of the most expensive, forms of energy today.

"I don't know if it will work or not," Berry responded. "The question is, 'Is it right?' I know it's right."

A short time later, word came that the governor would meet with the group.

"Remember to stand tough," Blanton reminded everyone. "It's not just about meeting with us, but meeting our demands."

Eighty-year-old Patty Wallace, who once thought she'd be making quilts at her age rather than participating in a protest, held up a sign that spelled out these demands:

1. Stop the destruction of land, water, and people by mountaintop removal.
2. Support the economic transition with good jobs for miners and communities.
3. Engage in sincere, civil, public conversation about how we solve these serious issues.

When the governor arrived, Berry spoke first. Several members of this group had repeatedly presented their concerns about the impact of mountaintop mining, he said, but they had never even received acknowledgement of a problem. "Instead, and far to the contrary, the government has publicly identified with the coal companies and has undertaken, with public funds, to support their interests in a court of law. We are here to say, as citizens and as taxpayers, that this is not acceptable."

Then, one by one, the others spoke up.

Bev May, a nurse from eastern Kentucky, showed the governor and a growing cadre of reporters a jar of brown water that she said came from the municipal water system near her home in Floyd County.

"Does that look like the state has done its job?" asked Stanley Sturgill, a retired coal miner and federal mine inspector from Lynch, Kentucky. "I worked in coal mines for forty-one years. Now, you know and I know and everybody here knows that mountaintop removal is a whole lot cheaper than coal mining. But it doesn't make sense when you end up with water like that."

Rick Handshoe, a retired police radio operator, spoke of there being so much methane in the water where he lived that he worried his elderly father's house might explode if he forgot to open the windows before showering. He spoke of the crawdads, frogs, and fish dying in the stream where he played as a child. "When chickens won't drink out of it," he said, "you know it's bad." And he invited the governor to come out and see it for himself.

After listening politely, with hands clasped in front of him, the governor said he could respect their differences of opinion; however, he thought that surface mining could be done responsibly. He then excused himself.

Blanton turned to the reporters and said, "We're not satisfied with the communication we had here today. We need clean water and a governor who will stand up for the people, not the polluters."

Berry added that he was pleased—because the exchange exposed the flaws in the governor's position. "He [the governor] thinks surface mining can be done without harm to land or streams. It's clear that nobody on our side thinks that's true, because we've seen the results or experienced them in our own families and homes. The idea here that there are two sides that can legitimately disagree is simply wrong. You can't rationally argue that the Earth ought to be destroyed."

And then they waited, expecting and hoping to be arrested, for the sake of the attention it would bring to the issue. But several hours later, the head of Capitol security said that they were welcome to stay through the weekend. Surprised, they decided to do so. A few people ran out to pick up pillows, toothbrushes, and sandwiches. Others sent out news releases and posted updates on Facebook. As the hours passed, they found places to read, talk, and sleep. At around 8:30 p.m., Berry stood up, put on his jacket, and headed for the door. Blanton called out in a nervous voice, "Wendell, Wendell, are you leaving us?"

"No," he said. "I'm going to brush my teeth."

On Monday, the group emerged from the building to the applause of about 1,000 people who had gathered outside for the I Love Mountains Rally, an annual event calling for clean water and energy. Despite the fact that things hadn't gone as they had planned, the group did attract attention for their cause in numerous publications, including the *New York Times* and *The Huffington Post*. They also succeeded in extracting a promise from the governor to come to eastern Kentucky and visit the mountaintop removal sites—a promise he fulfilled two months later.[19] And they deepened their commitment to putting the health of people and the planet first, no matter what.

"I don't think you should invest in things like this on the condition that you win," Berry said. "You have to do it for other reasons. You have to do it because you are committed to the health of the land and community and the people. You do it because it's right. And you have as much fun as you can."

THE INTELLIGENCE CONNECTION

The decision to engage in civil disobedience made by this group of Kentucky citizens (nearly one-third of whom were educators) was born out of an extended sense of empathy for the mountains, water, people, and other life—as well as an understanding of the interrelationship between human actions and natural systems. Each person had his or her own story, but perhaps Blanton's story best illustrates the effectiveness of leading with emotionally and socially engaged ecological intelligence—be it consciously cultivated or instinctive.

Blanton grew up on a dirt road in Harlan County, Kentucky. The daughter of a coal miner and the sister of a coal miner who died in a mining accident, Blanton left Harlan County briefly to marry, and then returned at twenty-five as a single mother of two. She bought a few acres of land behind her parents' house and settled into a mobile home with her children and two big dogs.

WHY CAN'T WE GO OUT AND PLAY, DADDY?

William "Bopper" Minton is a husky-voiced, grab-you-for-a-hug-on-first-meeting kind of guy. At forty, he lives with his longtime girlfriend and their seven-year-old daughter, Madison. It's not a big house, he explains, but it's his house—the same house he's lived in since the sixth grade. He remembers a childhood spent mostly outdoors, swimming, even bathing, in Little Goose Creek, which abuts his property. But when his daughter, who calls herself "Little Bopper," recently had her first sleepover, he would not let the kids go outside to play—just as he does not let her out of the house after she comes home from school. His eyes turn red, fill, and then overrun with tears as he tells the story.

"She kept saying, 'Why can't we go out and play, Daddy?'" It was a question presumably asked more for her friend's sake than her own, since she was already all too familiar with the answer.

When Madison was six months old and just beginning to crawl, Bopper gave her a bath, put her in a diaper, and took her outside. He gently placed her down on one side of the porch, walked to the other, and said, "Come to Daddy!" One can imagine the pride and joy—traces of which show on his face still—as he watched her crawl toward him. But when he scooped her up and held her in front of him, he discovered that his little girl was covered in black dust: hands, arms, belly, legs, feet, and toes. Having bathed her only moments before, he knew this could be only one thing: coal dust.

Furious, he picked her up again, put her in the car, and drove up the road to the big coal processing plant, which was built after his family moved into the home previously owned by his

Coal mining moved into her community of Dayhoit in the 1970s. Blanton didn't think much about it until 1982, when her children had to wait for the school bus on the same road coal trucks traveled many times a day, spewing coal dust stirred up by each set of passing wheels.

"You know, my son would say, 'Our shoes get dirty before we get to school,'" Blanton recalls. "And, one morning, he told me, 'Somebody has to drive us across this part of the road every day, because the coal muck is there.' I went down there, and I'd seen it," she says, referring to a large puddle of black water and coal sludge, "and it was just disgusting."

She called the county highway department to ask that they clean up the sludge, even by simply digging a ditch so the sludge would run off to the side of the road rather than pool near the school bus stop. Instead, the next day, a coal truck appeared outside her mobile home and circled it all day.

"I called the road department dude back," she recalls, "and said, 'By any chance, did you call the mining company?'"

When he said yes, she asked to speak to his supervisor, whose only response was, "Lady, you have to learn to live with this if you live in a coal mining town."

"I beg to differ," she said. "My kids aren't wading through muck to get on the school bus."

grandfather. Finding the best-dressed man working there, he said, "You look at my child. I just got her out of the bathtub. Now, you look at her legs."

About 100 coal trucks—then and now—pass Bopper's house every day, casting off coal dust and kicking up the dust on the road. He posts a "We Deserve Clean Air" sign in front of his house; when some of the drivers see that, he says, they deliberately swerve to the side to stir up more dust. Bopper's house, now black from coal dust, has lost most of its value because, as he says, "anybody would be crazy to buy it."

More significantly, his daughter has been diagnosed with severe sinusitis and allergies caused by coal dust. She is on eight medications—inhalers, eye drops, and pills—to help her breathe, plus a nebulizer (which administers medication in the form of a mist inhaled into the lungs) for times when breathing is especially tough. On the rare weekends when he breaks down and lets her play outside—though never in the creek—she must use the nebulizer by evening.

"I'd fight Godzilla over my child," says Bopper, an out-of-work mechanic from Manchester, Kentucky. And, in some ways, he has been doing just that for the past six years: asking, demanding, pleading that the coal company either move their plant or pay to move his family—something that he, a lifelong resident of this long-depressed region of Kentucky, cannot afford to do. Yet despite the difficulties he faces, he insists, he is not giving up.

Then she told the mining company that if they thought sending a coal truck to circle her home all day would intimidate her, they had another thing coming.

"I don't know why I felt so fearless," she notes. "I probably should have been afraid, you know, but I wasn't."

Blanton continued to contact the highway department, which eventually built a new road so her kids didn't have to catch the school bus on the same road traveled by coal trucks. But her problems were far from over. Her children began breaking out in rashes after taking baths, and she discovered that the groundwater feeding her well was poisoned with toxic chemicals from a nearby coal plant.[20]

Realizing that other people were also being affected, she joined forces with two other women to educate her community about the problem and asked state and federal authorities to test their water. The women were portrayed as hysterical housewives in the local newspaper, she recalls. But they succeeded in getting the water tested, which revealed that their wells had been contaminated with vinyl chloride, trichloroethylene, and other volatile organic contaminants for some twenty-five years.

In 1992, the EPA declared Dayhoit a Superfund site, giving the federal government the authority to clean it up. Some 5,000 pounds of contaminated soil were excavated (and trucked to Alabama to be stored next to a poor African-American community[21]). Next, the officials prepared to extract the contaminated ground-water through a pump-and-treat system. As Blanton investigated this process, she found reasons to fear that the carcinogens and other toxic chemicals would simply be transferred from the water to the air.

"In my mind, I knew that they were going to poison me and my kids all over again," recalls Blanton, herself a cancer survivor.[22] So she loaded her mobile home onto a flatbed truck and moved it, her children, and her dogs to the other end of the state.

Leaving Dayhoit, however, did not mean leaving behind her commitment to dealing with the problem of mountaintop and other forms of coal mining. She had come to understand all too well their impact on the people, land, and water. And she felt too much empathy for the people of her hometown—many of whom had died in their fifties and sixties.

Blanton became a leader in Kentuckians for the Commonwealth and founded the Canary Project, which focuses on reducing dependence on coal and other fossil fuels, strengthening mining laws and regulations, and demanding the enforcement of existing laws. She has spent thousands of hours lobbying in Frankfort, Kentucky, and Washington, D.C.

As she evolved into a leader who instinctively reflected the capacities of socially and emotionally engaged ecoliteracy, Blanton learned many lessons about effective leadership. Among the most essential, she says, are these:

> **Don't communicate from a place of anger.** "I started out expressing anger," she recalls. "I *was* angry. But that was not going to reach anybody." You must be aware of your own feelings, have the ability to control them, and develop the facility to interact effectively with others.

> **Reach people on the human level through stories.** "I could tell you one statistic after another," she says. "But that isn't going to reach you." Effective leaders have empathy toward others and make connections for people by focusing on the human impact.

> **Foster dialogue instead of debate.** "I always try to find a place in conversation where you and I agree on something," she says. "And once we agree on something, then we can go on to have a dialogue."

> **Speak from the heart.** Blanton learned this from Wallace, who told her at a rally years before, "Put down the damn speech and talk from your heart."

> **Make ecological connections clear to others.** As Blanton and others sang while waiting in the governor's office,

If you blow up the mountains,

Push it in the valley,

You gonna reap just what you sow.

In this story, Teri Blanton, Patty Wallace, Bev May, Stanley Sturgill, Wendell Berry, and others hold a deep connection to Appalachia, its people, and the landscape. Reflect on your own connection to a place, now or in the past. Can you think of a location for which you would be willing to fight? Think about your students. Do they seem to have a connection to the place in which they live? How might you help them develop such a connection?

Taking a Power Trip

Chapter
TWO

Although 45 percent of the U.S. population depends on coal power for electricity, the process used to extract coal is largely invisible to most people. In this story, Allyn Steele, a history teacher in South Carolina, makes the invisible visible by taking his students on a field trip to observe ecosystems before and after mountaintop removal and to meet people on both sides of the issue. As you read the story, notice how new knowledge and passion emerge as the process of mountaintop mining becomes increasingly perceivable to the students.

LIKE MANY COMMUNITIES throughout the United States, Spartanburg, South Carolina, seems like a place with few connections to coal. Situated at the foot of the Blue Ridge Mountains, this small college town boasts six institutions of higher learning and a vibrant arts community. Though Spartanburg was formerly a center of textile manufacturing, most of its residents now work in health care, government, or education. Others work at the corporate headquarters of Denny's or the nearby BMW manufacturing plant, the only one in the United States.

So when Allyn Steele, a history teacher at Spartanburg Day School, was challenged to inspire a sense of civic responsibility in his students, his decision to direct their attention to mountaintop mining in Appalachia was not an immediately obvious one—nor would it necessarily be a straightforward one to carry out.

He could, of course, point to some basic facts connecting the Carolinas to mountaintop mining. For example, Duke Energy, headquartered in North Carolina, is the nation's third largest user of coal acquired through mountaintop mining.[1] A Spartanburg Day School graduate and school board member ran a coal brokerage company. And trains carrying coal from Kentucky sometimes passed by the local greasy spoon, Ike's Korner Grille.

Like most of us, however, Spartanburg students didn't actually see coal being mined or burned. They couldn't truly grasp the meaning of "coal keeps the lights on." And they had little, if any, knowledge about or feeling for the impact of modern-day coal mining on the people and ecosystems of Appalachia.

So Steele designed a course called "The Power Trip," which examined the political, ecological, and economic consequences of coal energy—and most significantly, included a field trip to southeastern Kentucky.

BRINGING "THE ECOSPHERE" DOWN TO EARTH

The best way to inspire students to think about their relationship to the ecosphere is not by asking them to think about the ecosphere, author and farmer Wendell Berry argues, but instead to think about their local community.

In particular, he has proposed a curriculum that challenges young people to ask these eight questions about the places where they live[2]:

1. What has happened here?

2. What should have happened here?

3. What is here now? What is left of the original natural endowment of this place? What has been lost? What has been added?

4. What is the nature, or genius, of this place?

5. What will nature permit us to do here without permanent damage or loss?

6. What will nature help us to do here?

7. What can we do to mend the damage we have done?

8. What are the limits: Of the nature of this place? Of our own intelligence and ability?

RIDING THE COAL LINE

On an overcast Saturday morning in February, Steele, another adult chaperone, and six students set out to drive some 300 miles, tracking as closely as possible the coal line that ran between Spartanburg, South Carolina, and southeastern Kentucky. When they arrived nearly seven hours later, the director of the retreat center in Letcher County, where they would spend the next several nights, welcomed and warned them: "You can't drink the tap water. All the water in the water system is tainted."

Shortly after, they met with Patty Tarquino, an organizer with grassroots citizens' group Kentuckians for the Commonwealth, who would serve as their guide.[3] A native of Colombia, South America, home of Cerrejón, the largest open-pit coal mine in the world, Tarquino had come to Kentucky to help educate people about the hazards of mining. On the students' tour, she would lead them through what she described as a one-day, back-to-back blitz, packed with things to see and people to meet.

The first day began humbly with a hike in the Bad Branch Falls Nature Preserve. Owned in part by the Nature Conservancy, this preserve spans almost 3,000 acres and boasts the largest concentrations of rare and uncommon species in the state.[4] As the students made the trek, they occasionally helped each other scramble across a boulder or cross a creek, which Steele appreciated as team-building opportunities. But the primary goal of this trip was to see what Appalachia, one of the most biologically diverse regions in the country, looks like when undisturbed by human actions. Students saw rare flora and fauna and learned about how nature thrives in a system. They were told, as Steele recalls, "One has to be pretty respectful of all forms of life to keep that system in balance." They were also being prepared for a lesson in contrasts.

"This is a beautiful place for us to enjoy," Tarquino said as they reached a spot with a view of the 60-foot waterfall and pristine valley. "But a lot of valleys like this are being threatened right now."

"What's up with that?" a few students asked.

They were about to find out.

BEARING WITNESS

After the hike, Tarquino drove the students and some other guests further into the mountains, up a narrow country road traveled primarily by coal trucks. She explained that private coal companies had purchased this public land to mine the coal reserves below ground. As they drove, they passed several posted signs along the roadway:

"DYNAMITE"

"EXPLOSIVES"

"DON'T COME ANY FURTHER"

"I was kind of freaking out," recalls Steele. "One wrong turn, and we'd fall off the mountain."

Local residents Sam and Evelyn Gilbert, whose house had a view of Black Mountain, the highest mountain in Kentucky and one scarred by strip mines, also rode with the students that day. At night, they said, you could see the lights of the mining site from their driveway. And when a blast occurred, you could feel the whole house shake.

But what most concerned the Gilberts, they told the students, was that the mining company wanted to dump mining spoils into a stream that ran through their property. (The Gilberts fought the mining company's plan and eventually won, although they remain uncertain about the company's future actions.)[5]

Gilbert, who used to work at a strip mine, told the students that he knew what good surface mining was supposed to look like. He said that good surface mining practices required restoring a site to its original shape and soil contents. But that was not what the students witnessed. Nor is it what many people believe is possible after a mountaintop mining operation. After all, even if the land is restored to its former contours, the plants and animals that once thrived in that ecosystem often can no longer survive. The area looks oddly stark and lifeless as a result.

Moreover, once the mountaintops are blasted, the "millions of tons of crushed shale, sandstone, and coal detritus have to go somewhere, and the most convenient spots are nearby valleys," reported John McQuaid in *Yale Environment 360*. "Mining operations clear-cut the hillsides and literally 'fill' mountain hollows to the brim—and sometimes higher—with rocky debris." This leads not only to an altered landscape, but also to the burial of headwater streams and the permanent loss of ecosystems.[6]

In the final step before the coal is shipped, it is washed to remove soil, rocks, and other debris. This process produces massive amounts of coal slurry—water contaminated with toxins such as arsenic, lead, mercury, and selenium that is then stored in slurry ponds built into the side of the mountain.

The toxic water often leaks into groundwater supplies—and sometimes the pond walls simply break down. In 2008, for example, a Tennessee impoundment broke, releasing what the *New York Times* reported to be enough toxic sludge to flood more than 3,000 acres one foot deep.[7] When a West Virginia impoundment broke in 1972, it killed 125 people and destroyed thousands of homes.

"Meeting the Gilberts and seeing Black Mountain was just really, really sad," recalls college student Matt Roberts, who was fifteen years old when he made the trip. "It was just the worst case of environmental injustice I'd ever seen."

A BIRD'S-EYE VIEW

A few hours northwest of Black Mountain is Hazard, Kentucky, a city of nearly 5,000 people that President Bill Clinton once held up as an example of the failure of the war on poverty. It is a city surrounded by mountaintop mining, but its residents, many of whom have "Coal Keeps the Lights On" bumper stickers on their pickup trucks, cannot truly see the impact from where they live. To grasp the full magnitude of mountaintop mining in Kentucky, you need to get in a plane.

For that reason, on day two, Steele and his students headed to Hazard Airport, where Darwin Jones, a pilot from the conservation nonprofit organization SouthWings, took them up in a small plane that resembled "a Jeep Wrangler with wings," as Steele put it.

As someone who hates flying, Steele was more than a little on edge. But he got in first, followed by his students. What they saw together—for as far as the eye could see—was land utterly transformed from anything that looked natural.

"Words really can't describe the level of destruction and degradation and complete disrespect for the land there," said Roberts. "You go from gorgeous mountains and trees and valleys to sludge ponds and fractured earth."

After the plane landed, several students appeared to be in shock.

"I've never seen something so intense," one said.

"These were all mountains at one time?" another asked.

"How is that possible?" asked a third.

CREATIVE DISCOMFORT

On the last day of the trip, the students had many questions, and Steele created an opportunity for them to ask them at a coal processing plant.[8] But the encounter was far from a one-way street, as the company representative also took the opportunity to challenge the students.

"The representative said [the coal company] was doing the world a great service," recalls Roberts. "He said they left the land better than it was before. He said they were promoting economic growth [because you can build on flat land]. And, repeatedly, he said, 'Coal keeps the lights on. Are *you* going to stop using lights?'"

Most of the students did not take these statements at face value. They had already learned that mountaintop mining provides far fewer jobs than traditional mining ever did; sites are often too unstable to build upon after mining; and reclaimed land is never as good as the natural environment.

But Roberts was particularly moved to challenge what he heard when the coal representative stated, "There's nothing wrong with the water. I'm not worried about my water."

Having visited with Sam Gilbert, who lived near a creek as brown as chocolate milk, Roberts asked, "Are you on city water or well water?"

"I'm on city water," the representative replied.

"All the people we talked to are on well water," Roberts said, knowing that meant they were much more likely to be drinking water polluted by mining, since city water is treated to remove chemicals and impurities.

In the end, the coal representative stuck to his defense of mountaintop mining. But the students did not, as might be expected, adopt a simplistic oppositional posture that painted those who work in the coal industry as the "bad guys." What they learned instead was both more complex and realistic.

Roberts, now a neuroscience major at Rhodes College in Tennessee, said the experience taught him: "We're the ones to blame. We want a lot of energy, and we want it cheap. The coal companies are meeting the demands of the people.

It's a horrifying way of doing it. But if we weren't asking for it, they wouldn't be doing it. It's just the laws of supply and demand that make it cheaper and more profitable to blast the mountains."

Students recognized, Steele added, that "it's a systemic problem—we're all connected to this, we're all in on this."

And for anyone who might have missed the point, the connection was driven home to them as they left the plant, got into their cars, and headed back south to Spartanburg—in the same direction as a freshly loaded coal train.

WHAT'S MY CONNECTION?

"But what can we do?" a sixth-grade boy in Oakland, California, asked his teacher after learning about what mountaintop mining is doing to the drinking water, people, and environment in eastern Kentucky.

To a twelve-year-old who would soon be playing with his friends at recess, the day's lesson might otherwise have turned into just another dismissible tragedy—except for the fact that, with the help of an online application that uses Google Earth, he also saw the line that connects the electricity plant that powers his own school and home to coal from this region some 2,300 miles away.

As a teacher, here's how you can use the application, "What's My Connection to Mountaintop Removal?" (www.ilovemountains.org/my-connection): Ask students to type in their ZIP code. If your community's energy supply is connected to coal power and mountaintop mining, a red or black line will instantly appear on a Google Earth map, linking your community to the source of the coal. The line will be red if your energy provider uses mountaintop removal coal and black if it buys coal from companies that operate mountaintop removal mines in Central Appalachia. Students can then click on the mine or company name to learn more, or zoom in for aerial images that show where trees have been cleared, mountains stripped, and slurry ponds built.

Other features include a high-resolution tour of the stages of a mountaintop removal mine site, "before and after" images of twenty-two mountains destroyed by mountaintop mining, and videos of "America's Most Endangered Mountains."

BACK TO SCHOOL

Back in the classroom, Steele knew he had succeeded in awakening his students not only to the impact of mountaintop mining on the people and ecosystems of Appalachia, but also to the deep sense of caring that leads to action.

His next goal was to propel them into some kind of meaningful action. In his mind, that meant doing something more than declaring their commitment to being a more thoughtful consumer or writing about their experiences in a term paper. "The idea was to get them to reflect on their experiences and move those experiences out into the world," Steele said.[9]

The class discussed possibilities and invited Gary Henderson, a reporter with the *Spartanburg Herald-Journal,* to speak with them.

"Okay," Henderson said. "Who wants to help write a story for my paper?"

Christopher Riehle, now a student at the University of Chicago, volunteered. Together, Riehle and Henderson published an article in the local paper.

Other students made a poster about their trip and tried to serve glasses of water with lumps of coal inside at a public forum.

Roberts, Riehle, and students Stuart Long and Caroline Gieser also cofounded the first Spartanburg Day School Environmental Club, which grew from six to thirty members in two years.

THE LAST MOUNTAIN

The Last Mountain is a ninety-five-minute film that examines the issue of mountaintop removal coal mining. Used in conjunction with the Center for Ecoliteracy's *The Last Mountain Discussion Guide,* it offers a way to engage students in discussions about mountaintop mining.

Suitable for high schools, colleges, and community settings, the guide focuses on four themes suggested by the film:

- Everyone is connected to coal
- Private rights vs. the common good
- Democracy in the balance
- The value of a mountain

The guide offers suggested activities and discussion questions, such as the following:

1. Robert Kennedy Jr. asserts in the film that people should not have the right to destroy a mountain that they cannot re-create. Do you agree or disagree? Do you believe that humans have a right to use the Earth and its resources as they see fit, no matter the cost?

2. In what ways is the fight over Coal River Mountain (the last major intact mountain in West Virginia's Coal River Watershed) a fight about democracy?

3. Many companies and individuals are interested in making as much money as possible in the shortest amount of time. How does that compare with society's long-term interests? Is there a way to strike a balance between the two?

To download a free copy of *The Last Mountain Discussion Guide,* visit www.ecoliteracy.org/downloads/last-mountain.

Roberts, who served as president of the group for two years, went on to complete leadership training with the Sierra Club; he became the youngest member of the board of the Southern Energy Network, and served as head of the mountaintop removal task force for South Carolina's Palmetto Environmental Action Coalition and as president of Rhodes College's environmental group, Green Rhodes. While planning a career as a neurosurgeon, he said he has no doubt that he will continue to pursue environmental justice issues throughout his life—a commitment he credits directly to Steele and the class trip to Kentucky.

"Allyn created an experience that caught everybody's attention," recalled Spartanburg school headmaster Chris Dorrance. "There were people living here who didn't know much about Kentucky or really believe mountaintop mining was happening. And he brought back kids with passion, and that, of course, was very good."

Dorrance said that the experience also helped galvanize the school community about the larger issues of energy and sustainability, recognized by some as increasingly important on multiple fronts. For example, the late Roger Milliken, a prominent school board trustee and president of the world's largest textile and chemical manufacturing company, Milliken & Company, had recently told his fellow board members that rising oil prices and declining supplies made it imperative that they focus on sustainable practices—and the only way he would continue to make financial contributions to the school was if they did so.[10]

Dorrance, like many educators, has also observed students expressing more interest in sustainability in recent years. From the perspective of a school leader, he said, "I don't think there is any question that there is a growing imperative to address sustainability."

What matters most in the end for a school that values social, emotional, and ecological intelligence, he said, is not only exposing students to mountaintop mining or any of the other big environmental issues of our day, but also empowering them to do something about these real-world problems.

And both Dorrance and Steele agree that the key to success lies in making connections—between the lives of students and those of people and ecosystems at the other end of our energy, food, and water supplies.

"If we are to sustain ourselves, we have to think and act differently," Steele said. "And I think we can make education a valuable asset for that transformation."

Spartanburg students started out by visiting a nature preserve where they experienced the diversity of Appalachia's ecosystem undisturbed by mining. In what ways can educators increase students' direct experiences with nature so that they better understand the implications of habitat destruction?

PART TWO

FROM ANGER TO ACTION
IN OIL COUNTRY

The Heart of the Caribou

Chapter
THREE

Sarah James, Arctic Village, Alaska

This story reveals a way of life based upon a deep understanding of how nature sustains life, one of the five practices of socially and emotionally engaged ecoliteracy. Like many people in indigenous cultures, the people portrayed here reflect a profound connection with and reverence for other living beings, especially those with whom they share land. As you read, think about how nature serves as teacher in this indigenous culture and what—if we adopted a similar perspective—our society could learn about long-term sustainability.

D URING ONE OF THOSE LONG SUMMER days when the sun only briefly sets in Alaska, we traveled by boat from the southern edge of the Arctic National Wildlife Refuge (ANWR) along the Chandalar River, a loopy river system flanked by more mountains than even native Alaskans can name. We passed places where loons shared the river with beavers and muskrats, and you can catch whitefish with your bare hands when the water is high. Some hours later, with the massive Brooks Range in the distance, we tied our modest steel boat near a row of low glaciers and started a fire—making tea from river water, roasting smoked salmon on a stick, and feeling blissfully at one with nature. It seemed possible, even desirable, to stay in that place forever. But dark clouds over the mountains signaled our time to pack up and begin our journey back.

We took the turns of the river a good deal faster than we had on the way north in a futile effort to outpace the rain. When the rain came, it came down hard, and then harder—so hard, it hurt. We covered ourselves with a tarp and soldiered on for nearly an hour until Eddie Sam, the sixty-one-year-old sheep hunter at the helm, did what in this weather seemed the most unfathomable thing: He cut the engine and turned the boat around.

"We're going to climb a hill," explained Sarah James, an indigenous elder who grew up in this region and has traveled the world for the past twenty years to defend it. "He saw moose tracks."

Tying the boat to a tree, Sam grabbed his rifle and took the hill as if he were still an Army sergeant in Vietnam. From the peak, he scanned the landscape, but saw no sign of moose. Back in the boat, we continued down the river and soon spotted two moose standing on the bank, silently watching us.

"Want to shoot one?" Sam yelled over the roar of the engine.

But a quick look at his face revealed that he was joking. These moose were too young and lean to kill. Just two weeks before the summer solstice, the willow trees lining the Chandalar River were not yet green and full enough for moose to bulk up on, explained James, who was raised in a traditional subsistence lifestyle. It would not be right—nor worth the time and effort—to shoot the animals, skin them, carry them back to the village, butcher them, and share them.

We returned, therefore, wet and empty-handed, albeit content, to Arctic Village, home to Gwich'in natives who live at the southern edge of the ANWR, the largest wilderness area in the United States and one of the world's last ecosystems still relatively undisturbed by human impact. As the U.S. Fish and Wildlife Service notes, this area, about the size of the state of South Carolina, remains "pretty much as it has been since glaciers covered North America."[1]

Like the Inuit, the subsistence whalers who live on Alaska's North Slope, the Gwich'in are one of the northernmost indigenous tribes in America.[2] They are also among the oldest—believed to have survived above the Arctic Circle for the past 20,000 years. Unlike the Inuit, however, the Gwich'in have refused to benefit from oil development. To the contrary, they have spent the past several decades engaged in what writer Peter Matthiessen has described as "the longest and most critical environmental fight in our nation's history": opposition to proposed drilling for oil in the ANWR.[3]

More recently, the Gwich'in, who live in a place where temperatures range from 70 degrees Fahrenheit in the summer to 50 below zero in the winter, have also begun to address another oil-related issue: climate change, signs of which they see all around them.

After living in isolation for millennia, the Gwich'in decided to come forward to share a message steeped in traditional ecological knowledge and systems thinking. They perceive the natural environment as a place where everything is connected to everything else. They also perceive it as perfectly designed to support human survival—if we understand, respect, and relate to it on the basis of need, not greed, as James puts it. When we fail to do that, the Gwich'in say, nature and people suffer. The solution, however, does not lie in a return to past ways of doing things, but rather an integration of ecological intelligence into modern ways of living. And the Gwich'in believe that education is central to the discovery and development of these solutions.

Of the estimated 7,000 Gwich'in who live in fifteen villages across the United States and Canada, perhaps no one has done more to share the tribe's message and way of life than James.[4] At sixty-one, the Goldman Environmental Prize winner[5] has served as spokesperson for the Gwich'in for more than twenty years. From her remote and humble home, James has traveled widely—to Mexico, Guatemala, Brazil, Nicaragua, Ecuador, South Africa, the Netherlands, Denmark, and across the United States—to speak out against oil drilling in the Arctic. Although she describes herself as shy, James says she has risen to fulfill the public-educator role that her tribal leaders assigned to her, because protecting the Earth is her greatest duty. And to meet this challenge, she has brought the skills of socially and emotionally engaged ecoliteracy—learning to control her own emotions, relate effectively to others, and share her lifelong knowledge of our basic connection to the land, especially that unique part of it where she lives.

THE STORY OF OIL IN ALASKA

The story of oil in Alaska is really the story of three places, all of which lie on the state's North Slope at different stages of development. First, and most familiar, is the story of Prudhoe Bay, where a major oil field discovery in 1968 led to the construction of the 800-mile Alaskan pipeline. According to federal regulators, that pipeline is now corroded and poses serious public safety and environmental risks.[6]

Second is the story of the National Petroleum Reserve, a 23.5 million–acre expanse of public land to the east of Prudhoe Bay. Originally created in 1923 when the United States was converting its Navy fleet to run on oil, the Reserve remained essentially a wilderness area until the late 1990s, when 87 percent of it was opened to oil and gas leasing.[7] Since taking office, President Barack Obama repeatedly has supported onshore and offshore drilling for oil. And in 2011, the Interior Department gave conditional approval to Shell Oil Company to begin drilling in the Arctic Ocean as early as 2012.[8]

The third story is the most controversial and of greatest interest to the Gwich'in. In 1980, President Jimmy Carter more than doubled the wilderness protection area in the western part of Alaska to 19.6 million acres. In the process, however, he made what has been widely regarded as a concession to oil developers by setting aside a 1.5 million–acre tract of land for fossil fuel exploration. Known as the Coastal Plain, or Section 1002, these 1.5 million acres have been the source of conflict ever since.

The debate has focused on just how much oil exists in this area and what impact drilling would have on the resident wildlife—above all, the Porcupine caribou, which have been an integral part of Gwich'in culture for as long as anyone remembers.[11]

TOP TEN OIL PRODUCERS

Here—out of order—are the top ten oil-producing countries in the world. Test your knowledge by ranking them from one to ten. Then check your answers in the note below.

Brazil   Mexico

Canada   Russia

China   Saudi Arabia

Iran   United Arab Emirates

Nigeria   United States

(The top ten producers, in order, are Saudi Arabia, Russia, United States, China, Iran, Canada, Mexico, United Arab Emirates, Brazil, and Nigeria.[9] In the United States, here are the top six oil-producing states, in order: Texas, Alaska, California, North Dakota, New Mexico, Oklahoma.[10])

CARIBOU PEOPLE

Every spring, the 123,000 Porcupine caribou that live in the ANWR make the longest migration of any land animal on Earth.[12] Traveling from the southern end of the ANWR to the north, they cross rivers; battle fox, bear, lynx, and wolverine; and cover wide swaths of tundra in search of lichen, a plant that, when eaten, expands in their stomachs and helps sustain them over great distances—sometimes up to 3,000 miles along a distinctly less-than-direct path—to reach their destination. During this period of migration, the Gwich'in, who have hunted caribou for millennia, never shoot them, even if they spot the animals just on the other side of the river.

The Gwich'in are, in fact, so closely identified with the caribou that "Gwich'in" means "caribou people." Caribou meat has long been a central part of their diet, and the Gwich'in living in Arctic Village say they just don't feel full unless they eat it. The meat, they say, makes them strong. In generations past, Gwich'in also used other parts of caribou to make clothing, shelter, and tools. On a more spiritual or mythological level, Gwich'in believe that their origins can be traced back to a heart shared with caribou, which enables each to know what the other is doing.[13] And the Coastal Plain, the contested 1.5 million acres in the north, is the place where caribou life begins—a traditional birthing ground that is considered sacred land.

It is to the Coastal Plain that the caribou travel every year to give birth to their calves. They do this because it is a safe place with relatively few predators, and the winds blowing off the Arctic Ocean keep the mosquitoes away—a fact of no small consequence, given that mosquitoes prey fiercely on caribou, especially when the caribou are young and their fur has not yet grown as thick as an adult's.

"We want people to leave that place alone so the caribou can always come back and stay healthy," said James. "We believe God put us where we are today to take care of this part of the world. And caribou are one of the animals we try to protect. But it is not the only one. We are talking about all life. We depend on birds and ducks, fish, small animals, some plants, the water, caribou, moose, sheep. They're all important."

Moreover, since controversy over drilling in the Refuge developed, there has been a new awakening among the Gwich'in Nation that they hope will inspire others. They say that taking care of the Earth is not a role they have assumed as activists, but rather something that reflects their millennia-old way of life. It is also, they add, a way of life that is instinctive to children if adults do not get in the way.

THE GWICH'IN WAY OF LIFE

Archaeological evidence suggests that the Gwich'in may have lived in the area near Arctic Village for as long as 6,500 years.[14] In his book *Caribou Rising*, Rick Bass calls the Gwich'in "the most native of native peoples, nearly geological in their integrity, their endurance. Not as ancient as the oil we seek to burn in our course, but, still, more ancient than anything of mankind left on this green Earth."[15]

Throughout most of their history, the Gwich'in led a highly nomadic way of life; James notes that they were strong and healthy, living off the land and traveling long distances. But when they came in contact with European fur traders in the 1800s, the Gwich'in encountered alcohol, foreign diseases, and other challenges to their traditional existence. By the 1900s, they began settling in villages, and by the 1950s, most had given up their nomadic practices.[16]

About 160 Gwich'in now live in Arctic Village. There is no police department, fire department, doctor, dentist, or household plumbing. No roads lead in or out. Contact with the outside world comes largely by way of a bush plane that lands at the reservation twice a day, bringing supplies, mail, and occasional visitors. When someone needs medical attention, he or she must fly 300 miles to Fairbanks.

Only a handful of jobs are available through the market, post office, and leadership committee. Many people rely on a yearly stipend from interest on the state's permanent fund, created in exchange for leasing the oil drilling rights in the North Slope through which the Alaskan pipeline runs. This is not enough, however, to elevate the people out of poverty. The median household income is $22,500, with almost 35 percent of the population living below the poverty level.[17]

As on many reservations, contradictions between the traditional values of the people and the realities of modern life are also visible. Broken-down bikes and ATVs are left discarded on people's property. Many young people pass noontime hours asleep on the couch in front of a TV. And a significant number of residents are obese, a result of an increasingly unhealthy lifestyle and the consumption of processed food sold in the village market. Residents also report that drug abuse is not uncommon.

And yet, the Gwich'in of Arctic Village are one of only two villages of indigenous Alaskans who refused to accept oil development and pipeline money when the Alaska Native Claims Settlement Act of 1971 granted $1 billion and 44 million acres to Native Americans. Instead, they held out for the original tribal land claim of 1.8 million acres, "in the full knowledge," as Matthiessen has written, "that this brave commitment to the integrity of their ancestral land and their traditional ways would condemn them to a life of bare subsistence."[18]

THE PROSPECT OF DRILLING

In the 1960s, a young geologist for British Petroleum (now BP) hiked across the Coastal Plain of the ANWR and observed rock formations indicating the potential for vast quantities of oil.[19] Roughly a decade later, that finding ignited the first of many fights over drilling, as environmentalists sought protection for the Refuge. It is also what led President Carter to the ill-fated compromise that doubled the size of the Reserve but left open the possibility of drilling in the Coastal Plain. No serious threat developed, however, until 1987, when the Reagan administration's Interior Department recommended that Congress allow drilling there.[20]

In response, the Gwich'in did something they hadn't done since World War II: They organized a gathering of all Gwich'in tribes from fifteen villages in Alaska and Canada. From June 5 to 10, 1988, they came together for a meeting in the Arctic Village community hall. The elders wanted the whole nation to discuss several issues: maintaining the Gwich'in language, border relations, the problem of drug and alcohol use, and, of course, the caribou.

At the beginning of the meeting, one of the elders pulled out a staff topped with an eagle head and explained that it would serve as their talking stick. No agenda, note taking, speaking in English, or outside visitors were allowed. Everybody had a chance to hold the stick and speak.

It was decided that the only way the world would know about them and their reliance on caribou was if the information was presented in black and white. So the village chiefs left the community hall, climbed a hill, started a campfire, and wrote a resolution advocating for the preservation of what they identified as the calving and post-calving grounds of the Porcupine caribou herd.

The resolution declared that the caribou were essential to meet the nutritional, cultural, and spiritual needs of the Gwich'in people; that the Gwich'in have the inherent right to continue their own way of life (as affirmed by international human rights covenants); and that the caribou are endangered by proposed oil and gas exploration and development. It called on the U.S. Congress and the President to recognize the rights of the Gwich'in people by prohibiting development in the Coastal Plain and protecting it as a designated wilderness area.[21]

After returning to the community hall, the chiefs presented the resolution to the elders, who voted to pass it. They then announced that they would need to select a group of people to implement the resolution. They chose four representatives from Canada and four from the United States to form the Gwich'in Steering Committee. Sarah James was named chairperson.

GIVING VOICE TO A WAY OF LIFE

James has long gray hair and an easy smile. She lives in a roughly 20-by-20-foot cabin with an outhouse and smoking hut out back. Her couch serves as her bed, and she eats and works on a solitary table. On her bookcase are a number of bumper stickers, including one that offers her response to former Alaska Governor Sarah Palin's "Drill, baby, drill!": "Chill, baby, chill."

The youngest of nine children, James grew up learning subsistence firsthand from her parents, who built their own cabin, made their own clothes, and caught, killed, or harvested their own food. She loved their traditional way of life and hated it when she was sent off to boarding school at the insistence of the Bureau of Indian Affairs (BIA). She recalls that she was a poor student—both literally and figuratively. At thirteen, she was reading at second-grade level. With little money and borrowed clothes, she eagerly awaited packages from her parents, who sent cranberries, salmon strips, and an occasional $5 bill.

"Every time, I would write and say, 'I want to go home,'" she says. "And each time, they would write back and lecture me: 'The only way you are going to make it in the world is education.'"

So she stuck with it and squeaked by with a C-minus average. The BIA later relocated her to San Francisco, where she learned typing at a business college and was briefly employed by an insurance company. But when she was called home for her father's funeral, there was no money for her to return to San Francisco. She stayed in Arctic Village and tried to start a school. She gave birth to a son and weathered a failed marriage. She worked as a healthcare aide for more than a dozen years but found it unfulfilling.

As it turned out, she did not quite discover her place until the elders discovered it for her, giving her the responsibility of advancing their mission of protecting the caribou, the Coastal Plain, and the Gwich'in way of life. On the surface, she may have seemed an unlikely choice, as she describes herself as shy and less than adept at writing or public speaking. "I'm a very nervous person," she says. "When I have to talk to people, my heart beats a lot. I sweat. I'm a nervous wreck. I'm afraid I'm going to miss something important to say or that I won't be clear. I always pray a lot beforehand."

But she possessed several other qualities central to socially and emotionally engaged ecoliteracy, including a deep ecological knowledge gained from her parents and other village elders, a love of the Earth, a friendly nature, a willingness to share her story, and a quiet but firm determination to do what she thinks is right for the world.

"I believe that creation is perfect the way it is, and everything is there for a reason," she adds. "If we could just see it and follow it and use it that way, then things would be in balance."

REACHING OUT

James and her fellow members of the Gwich'in Steering Committee made sharing the Gwich'in story with other Americans their top priority. They began reaching out to newspapers and magazines to tell the tale of an ancient people trying to protect their way of life from oil development. The Gwich'in soon succeeded in attracting attention from *National Geographic, People* magazine, *The Christian Science Monitor*, and other high-profile publications. They enlisted the services of the Public Media Center, a San Francisco-based strategic communications organization, and helped raise funds to post billboards in Washington, D.C., and elsewhere. James traveled widely to speak at almost any event where people would pay her way. She focused her efforts on educating members of Congress about the conflict between oil drilling and the traditional way of life. ("They need education," she quips. "They need a lot of people to educate them.")

Along the way, she relied on a capacity for ecological intelligence that she had cultivated since birth, and she embraced numerous opportunities to develop the emotional and social intelligence she needed to engage effectively in this kind of educational leadership, including the ability to manage her own fear, anxiety, and anger and to communicate well with others, especially those with whom she disagreed.

For example, she once seized a chance to meet Alaska Senator Lisa Murkowski at a reception. Murkowski has long been supportive of drilling in the Refuge—so supportive, James assessed, that challenging Murkowski directly would be doomed to fail. So she started from a point of agreement.

"I introduced myself and thanked her for her good work on women's issues. I said we should do more about women's issues because women are sacred. They give life," James said. Then she made a tactical transition, telling Murkowski that she thought of the Coastal Plain as a women's issue. "It is a birthplace and nursing ground, and life is life."

On another occasion, she met with the late Ted Stevens, then Senator from Alaska and a staunch supporter of drilling. As soon as she began speaking, James recalled, Stevens interrupted her and yelled, "I have something to say, too!" He then lectured her about the need to drill for oil.

Rather than get angry or give in to intimidation, James sat with her hands in her lap, waiting for him to finish. Then she said, "We are caribou people. It makes us who we are, and we must protect the Coastal Plain."

"It's not going to happen," Stevens said. "I'm done."

Despite Stevens' refusal to listen, James said she would not allow herself to get angry. "It is worth it to talk for the Earth. That's all that is important to me. I don't waste my time or energy on what I cannot change. But I still come in and say my part because I have a right to. If I get intimidated, they will work on me more to get me angry, and it will not turn out to be a good message." And how the message is communicated is as important as the content of the message itself, she says. "As the elders told us at the gathering in 1988, 'We need to educate the world about why we say no to oil and do it in a good way.'"

IT'S NOT OVER TILL...

For James, educating in "a good way" means speaking the truth respectfully and putting the Earth first. While undoubtedly an honorable approach, it is not necessarily always an effective one. In fact, it appeared to be doomed to fail in March 1989, as the U.S. Senate Committee on Energy and Natural Resources approved leasing in the Coastal Plain. Drilling seemed inevitable.

Just one week later, Captain John Hazelwood ran the *Exxon Valdez* oil tanker aground on a reef, spilling nearly eleven million gallons into Prince William Sound.[22] With more than 1,000 miles of Alaska's coast severely damaged, it was considered the largest spill ever in American waters—until the 2010 Deepwater Horizon spill in the Gulf of Mexico.[23]

President George H. W. Bush announced that he saw "no connection" between the *Exxon Valdez* accident and drilling in the Arctic. But the public and political mood shifted all the same, and the push for drilling came to an abrupt halt—at least for a few years. It has since resurfaced several times, especially after President Bush made offshore drilling a part of his energy policy.

Since taking office, President Obama has repeatedly supported drilling in the Arctic, both onshore and offshore, although he is opposed to drilling in the Refuge.[24] Still, the Gwich'in and others called on Obama to give the Refuge national monument status in honor of its fiftieth anniversary in 2010—the only

---

UNITED STATES IS NUMBER ONE IN OIL CONSUMPTION

The United States consumes more oil today than any other country in the world. At 18.8 million barrels of oil a day, as of 2009, the United Sates consumes more than twice as much as China, which has four times as many people. More than half of the oil Americans use fuels cars and trucks—as a result of low fuel-efficiency standards, the longest commutes of any industrialized country, and an American love affair with cars.

Here are the top ten oil consumers (barrels consumed per day) in 2009[25]:

United States	18.8 million	Germany	2.5 million
China	8.5 million	Saudi Arabia	2.5 million
Japan	4.4 million	Brazil	2.4 million
India	3.0 million	South Korea	2.2 million
Russia	2.9 million	Canada	2.2 million

move they believe would finally put the contested Coastal Plain out of reach of oil developers. As of yet, however, that call has not been answered.

ON THE FUTURE AND EDUCATION

Although James and the Gwich'in have made protecting the ANWR their top priority, they also see this issue within a broader ecological context and share the view that we are approaching a critical turning point, with human actions threatening many of the natural systems on which we depend.

"We have always spoken for the Earth—always," says James. "Our elders from way back gave us a good message. People just don't listen. Now we believe it's going to get a lot worse. People are going to get hungry. There is going to be chaos. But people here and in other places might survive longer; if they go back to the land, they might survive a little longer."

More constructively, James says, people can stem the current tide of ecological destruction by changing their ways of thinking and behaving. And to that end, she says, "Education is central." Education, in fact, offers the opportunity to build on children's inherent love of nature and, from there, to cultivate a deep capacity for emotional, social, and ecological intelligence.

Education also offers the opportunity to teach children and adults concrete practices that can make a positive difference today. For example, James recommends reducing reliance on oil in many ways, including driving less and biking and walking more, allowing the principle of "need not greed" to govern consumption decisions, and not only recycling but also reusing. "There's nothing hard about it and nothing easy about it," she says. "It's just something you do every day. And if everybody does it, it will become a way of life."

Fundamentally changing our lifestyle to reflect a new ecological sensibility for the twenty-first century is, of course, no simple feat. The challenges before us are momentous.

But the Gwich'in, for one, will not be backing down, says James. "To protect the Earth is our way of life. It makes us who we are. There is no way we can walk away from our way of life, regardless of hard times. We just can't change and be someone we're not."

Reflecting on the ways Sarah James and her Gwich'in community embody socially and emotionally engaged ecoliteracy consider: How might you integrate some of these attributes and behaviors into your life? And how might you nurture them in your students?

Beyond Whining

Kids Rethink New Orleans Schools

In this story, students in New Orleans, Louisiana, face repeated unanticipated disasters yet find ways to be part of the solution by helping to build a more resilient city and school system. It is a journey that fosters socially and emotionally engaged ecoliteracy every step of the way. As you read, notice the multitude of unintended consequences resulting from human practices that exacerbated the impact of Hurricane Katrina and the Gulf oil spill, and the ways in which these students are helping build a more resilient New Orleans.

NEW ORLEANS IS A PLACE OF many distinctions. It survived one of the worst hurricanes, and the costliest natural disaster, in American history. It was hugely impacted by the worst accidental oil spill in the history of the oil industry. It has the highest per-capita murder rate in the nation. And, for decades, its schools were ranked among the worst in the United States. But in recent years, it has begun to develop another kind of distinction, as a place where middle and high school students, guided by some particularly wise and caring adults, are cultivating a culture of innovation across a broad range of increasingly important issues, including how schools use oil.

The students call themselves the "Rethinkers" and have proven to be unusually effective. The work of "Rethink: Kids Rethink New Orleans Schools" has attracted coverage from *The Times-Picayune, The Christian Science Monitor,* ABC News, and *The Huffington Post,* to name a few. The Rethinkers are one of the few student groups in the country to have signed a contract directly with a major food service supplier (ARAMARK) to serve locally grown produce twice a week.[1] School officials have adopted many of their recommendations, such as adding school gardens and installing hand-washing facilities in cafeterias. And they have gained the respect and support of numerous community and education

55

leaders, including Paul Vallas, the first superintendent of the Louisiana Recovery School District.

"Paul is obsessed with the Rethinkers and wants Rethinkers clubs in all schools," says Siona LaFrance, the superintendent's chief of staff. "He likes that the kids are thinking and challenging authority, and that all of their suggestions are based on a lot of consideration. And he likes that this is a continuing effort."[2]

Rethinkers founder and executive director, Jane Wholey, who instinctively grounds her work in socially and emotionally engaged ecoliteracy, wants to help students have a voice in how schools operate and to contribute meaningful recommendations for reform. She believes that with proper guidance, their participation can be both a learning experience and deeply empowering.[3]

"I tell kids there is no way you will ever be heard if you whine," says Wholey. "The ultimate goal for the Rethinkers is to identify a problem and solution or recommendation for change. We start with something they identify in their schools that makes them unhappy."

SPARKED BY TRAGEDY

The Rethinkers came together in 2006 as a result of Hurricane Katrina, which caused flooding in 80 percent of the city of New Orleans, nearly 2,000 deaths, and an estimated $81 billion in property damage.[4] It was, of course, the biggest event to rock their young lives. Of the city's 126 public schools, 110 were destroyed. An estimated 50 percent of the population fled the city.

Some of the students who left the city with their families enrolled temporarily in other schools for what Wholey describes as an "enforced year abroad." It was "a unique and sometimes traumatizing moment in the lives of young New Orleanians," she says. "But for many, it had also been a first glimpse of schooling far better than they had known in the notoriously dysfunctional New Orleans system."[5]

Dudley Grady Jr., now a student at Xavier University of Louisiana and a Rethinkers intern, remembers it well. "The bathrooms were the biggest things for me," he says. "To see a clean restroom in school? I'd never seen that before. Toilet paper? Soap? Mirrors on the wall that were not broken? I'd never seen that." Attending schools outside New Orleans, he was equally surprised to see doors on the bathroom stalls, enough desks and chairs for every student, and books that he could take home to read.

Before Katrina, two-thirds of New Orleans schools were failing and chronically under-enrolled. Some were so old and neglected that school board members said they should be condemned. Nearly three-quarters of eighth graders lacked proficiency in math or English. And the school system had a reputation for fraud, incompetence, and corruption.

With most school buildings destroyed and half the residents having left New Orleans after Katrina, there was widespread agreement that restoring the city's schools was an essential part of restoring the community. As a result, state legislators brought some 90 percent of New Orleans schools under the control of the Louisiana Recovery School District, a reform effort designed to transform underperforming schools into successful places for children to learn.[6] "We used Katrina as an opportunity to build—not rebuild, but build—a new school system," said superintendent Vallas.[7] This massive effort has been described as the "most ambitious system-wide reform in U.S. education history."[8]

The success of any reform effort depends on many things. But one key element is school culture and the ways in which it transmits (or fails to transmit) new values and behaviors. "Understanding school culture is an essential factor in any reform initiative," wrote Elizabeth R. Hinde, associate professor of Teacher Preparation at Arizona State University's Mary Lou Fulton Teachers College. "Any type of change introduced to schools is often met with resistance and is doomed to failure as a result of the reform being counter to this nebulous, yet all encompassing facet—school culture."[9]

It was in this spirit that Wholey believed that students, one of the central networks of relationships within any school system, should play a part in reforming their own schools. Driven by an instinct for creating a culture of trust and respect, she sought to engage students in the process of forging new norms.

"We know the issues," says Wholey. "There are not 200 of them. There are a dozen or so—in New Orleans and all over." The issues included school safety, cleanliness and bathrooms, availability of supplies and books, quality of food and cafeterias, quality of teachers and teaching, extracurricular activities, dignity, and, more recently, the use of energy.

Since 2006, the Rethinkers have met in clubs throughout the school year and for a six-week training program in the summer. They use "circles," a variety of exercises, and presentations by outside professionals to explore the issues. The goal is to create a safe environment for students to experiment and to encourage—even reward—innovation and experimentation. In short, Wholey holds the space for something new to emerge.

"The trick is always to allow it to be their show, their story," Wholey adds. "We could invite the most significant engineers to come up with solutions. But we've discovered kids almost always have things to say that are different from adults."

OPENING DOORS

The Rethinkers held their first news conference in front of Sherwood Forest Elementary School on July 21, 2006. It began with students Melissa and Amber Augustine opening its doors, which had been shut since the storm. The stench of decay poured out, overwhelming those who had come to observe. Pieces of desks, chairs, and a piano could be seen piled in a heap. The sisters then walked over to two desks left empty in honor of their cousins who died in the storm. Quietly, they lit a candle on each.

Sixteen-year-old Shannon Taylor described what New Orleans students had "endured," even before Katrina. She spoke of one student forced to sit on the edge of a table every day because there were not enough chairs in the classroom. Another student never owned a knapsack—and never needed one, because students were not permitted to bring books home. She described the filthy and broken bathrooms, lunches consisting of greasy hamburgers and donuts, and students who were unprepared for the Louisiana Educational Assessment Program (LEAP) tests that determine whether a student will advance to the next grade. "No wonder so many of us failed," she said.

Moving from problem to solution, the students offered a series of recommendations that included clean bathrooms, books of their own, safe schools, teachers who love to teach, and adequate extracurricular activities.

Shortly after the news conference, the Rethinkers also followed up with a report on students' perceptions of post-Katrina public schools, based on a survey of more than 500 students.[10] Among their findings were the following:

- Students felt least safe and secure in schools with security guards.
- Students gave the lowest possible rating to school food and the second-lowest rating to cafeterias and extracurricular activities.
- Only 62 percent of students, on average, felt that teachers were dedicated to and liked teaching children.

In 2007, the Rethinkers offered specific recommendations for improving school bathrooms. They tackled cafeteria food the following year, recommending the integration of local, garden-fresh ingredients; the elimination of "sporks," which they deemed undignified and impossible to use, in favor of real silverware; and the installation of sinks in cafeterias so they could wash their hands before eating. Turning to issues of safety and dignity in 2009, they recommended that metal detectors be replaced with "mood detectors" (students assigned to observe students as they arrived in the morning), a "chill-out zone," and a resolution circle.

Since then, says LaFrance, the district has adopted a number of the Rethinkers' recommendations: Sporks, for example, have been banned and replaced with silverware. Hand-washing stations have been installed in cafeterias. And a new agreement has been signed with the food service provider to offer more locally sourced food.

AND THEN CAME THE SPILL

When the Deepwater Horizon offshore drilling rig exploded in the Gulf of Mexico five years after Katrina, the magnitude of the disaster was not as immediately obvious to residents as that of the hurricane: Houses were not under water. No one could see the oil, the helicopters, or the cleanup crews from home. And most people did not lose family members. But during the second month of the spill, its magnitude had begun to become clear. "We know this is as big and bigger than Katrina," said Angelamia Bachemin, a musician who had stayed through Katrina and now worked with the Rethinkers. "We know this will affect our lives forever," added Dudley Grady Jr.

The Gulf of Mexico is one of the most diverse ecosystems in the hemisphere, a stopping point for migratory birds from South America to the Arctic, and home to abundant wildlife and natural resources. But it also bears the striking consequences, as the *New York Times* put it, of our "economic pursuits and appetites."[11] About 90 percent of the country's offshore drilling takes place in the central and western parts of the Gulf of Mexico, with 4,000 offshore oil and gas platforms and tens of thousands of miles of pipeline. Before the Deepwater Horizon explosion, an estimated half a million barrels of oil and drilling fluids had been spilled offshore. Even more had been spilled from pipelines, vessel traffic, and wells.[12]

But when the Horizon exploded on April 20, 2010, just 130 miles from New Orleans, it brought these issues to an unprecedented boiling point. The 185-million-gallon spill continued for three months before being capped, with oil covering more than 170 miles of shoreline from Louisiana to Florida. Eleven lives were lost, along with the livelihoods of countless more. The incident raised pressing questions about Americans' dependence on oil, as well as the long-term economic viability of New Orleans.

THE WORLD'S MOST IMPORTANT ENERGY SOURCE

Petroleum is arguably the natural resource on which our global economy most depends. Or as Michael Wilson, a research scientist at the University of California Berkeley, has put it: Oil is the "material basis" of modern society.[13]

Oil is formed from the remains of animals and plants that lived in the ocean some 300 to 400 million years ago. Remains were covered by layers of sand and silt that exerted great heat and pressure on them, eventually transforming the remains into crude oil and gas.[14]

People long ago realized that the smelly liquid they found in puddles on the Earth's surface could be used to keep lamps burning. But in the mid-1800s, two discoveries unleashed an explosive century and a half of oil development and economic growth: one was that people could access the giant oil reserves underground through drilling; the other was that they could separate oil into a range of petrochemicals in what are now known as oil refineries—and that these chemicals, in turn, could be used to make a great number of things, from gasoline to plastics.

Oil is now the world's most important source of energy, accounting for 35.9 percent of the world's total energy consumption in 2006, according to the U.S. Energy Information Administration. (Coal ranked second at 27.4 percent, and natural gas was third at 22.8 percent.)[15]

Once you start looking, it's difficult to find things that are *not* dependent on petrochemicals today. Among the products made from petrochemicals are appliances, cleaning supplies, clothing, computers, emergency and surgical equipment, fertilizers, furniture, heating and cooling systems, medicines including aspirin, phones, sports equipment, toothpaste, and vitamins.

For the Rethinkers, who had come together for their six-week summer program at the time of the spill, it also created a new focus: What could they do about it? To support the students' inquiry, Wholey and other leaders organized a series of events to shed light on the issue of oil dependence. Betty Burkes, an educator and activist who has worked with the United Nations and serves as the Rethinkers' curriculum developer, led one of the first events.

"As a teacher, one is always inquiring and looking for ways to make learning real," says Burkes. "So I decided it would be a good idea for them to get out of their seats, walk around with a paper and pencil, notice what is in the room, and imagine where it is coming from. In doing that, they discovered how dependent their school life is on oil."

Discussions about how schools might reduce their dependence on oil followed. This exercise helped make the issue not only personally relevant to students but also an experience that could lead to lasting change.

"I think to the extent that we can make learning personal, it has sustainability. To the extent that it isn't personal and remains an abstraction, it isn't sustainable," says Burkes. "And I think emotional, social, and ecological intelligence really helps to make learning personal, relevant, and sustaining."

FROM OBJECTS TO RELATIONSHIPS

As the Rethinkers continued their study of oil—in the midst of the three-month-long spill—several students went to the Crescent City Farmers Market on St. Charles Avenue to talk to shrimpers about how the spill was affecting them. Kay Brandhurst, whose family has been in the fishing (and, previously, whaling) business for generations, told the students she was still catching shrimp in Lake Pontchartrain, the second-largest saltwater lake in the United States. "But the oil spill makes me nervous for future seasons," she said.

"How can we help?" one student asked.

"Lots of prayers," she said.

Shortly thereafter, the Rethinkers staged a mock trial in which they asked another more complex question: Who is really responsible for the oil spill in the Gulf—BP or everyone who uses oil?

Responsibility for any spill is, of course, not a simple either-or proposition. A *Wall Street Journal* investigation, for example, found that the designers and operators of the Deepwater Horizon were to blame for shortcuts, lack of preparedness, and errors in judgment that likely caused the explosion.[16] And the U.S. government, the newspaper later reported, was responsible for requiring BP to base its preparations on outdated spill models.[17]

But the Rethinkers concluded that the American people ultimately bear the primary responsibility for the spill and the resulting damage. As one student argued, oil companies are a lot like gun manufacturers. When someone is shot and killed, we don't hold the person who made the gun responsible but rather the person who used it.

FOCUSING ON SOLUTIONS

Because one of the guiding principles of the Rethinkers is to be constructive, the identification of this problem had to be followed by a proposed solution, which they presented at a news conference on July 15, 2010 (coincidentally, the same day the spill was finally capped).

"If we want to prevent another oil spill," said ninth-grader Danny Do, the son of a shrimper, "we need to start weaning ourselves off this product and begin searching for new ideas. Now is the perfect time to get moving, and schools are a great place to start!"[18]

New Orleans schools, the students announced, should take steps toward becoming oil-free by 2015.

"We know that 'oil-free schools' sounds easy to dismiss, because it's such a big vision," says Mallory Falk, a recent Middlebury College graduate and community organizer who works with the Rethinkers. "That is why our focus now is to come up with realistic, practical ways for schools to move toward being oil-free."

In 2010, the Rethinkers offered four suggestions to schools:

1. Start measuring energy waste.
2. Form student green teams to identify ways to reduce waste and persuade other students to get with the program.
3. Eliminate the use of incandescent light bulbs.
4. Recycle materials instead of discarding them.

A year later, it was clear that these efforts had firmly taken root. For example, says Burkes, the Rethinkers set a goal of having only one garbage bag left for disposal after a retreat with twenty-three kids. And they did it.

She adds that the students' ongoing efforts to push for more locally sourced food and school gardens is also connected to their oil-free resolution, reflecting their growing awareness of the interconnectedness of all these issues. "I'm not sure we thought about it that way intentionally," she says, "but we came to see that each part had influenced the other part."

Whether New Orleans schools will actually become oil-free by 2015 remains to be seen, of course. But, says Wholey, the point is that any effort toward an admirable goal is worthwhile.

"This was the great lesson of our first year after Katrina: It is, in fact, therapeutic to get people to act to right the wrong done to them," she says. "It can be big or little. But the act itself is therapeutic. Pushing to make schools oil-free is our way of contributing."

"I'm interested in kids being grounded in inquiry and not blame, in being curious and not victims," says Betty Burkes, an educator and activist who has worked with the United Nations and serves as the Rethinkers' curriculum developer.

In the midst of the BP oil spill that leaked 185 million gallons of oil into the Gulf of Mexico, Burkes asked the New Orleans Rethinkers to identify all the ways in which their schools used oil—and then gather in a circle to brainstorm how they might use less.

As expected, the students came up with a long list of items made from petroleum, including pens and glue, desks and chairs—almost everything—recalls Jane Wholey, the Rethinkers founder and executive director. But they also focused on several school systems that were largely dependent on oil.

"A big piece of it was transportation to school," says Burkes. "One of the decisions that charter schools made in New Orleans was to get rid of neighborhood schools. So almost every kid is bused. The students realized in the circle conversation that to get rid of buses, they would have to revert to neighborhood schools." But then they had another idea: Why couldn't public transportation be solar-powered in a city with so much sun?

The students also discussed the use of air conditioning, especially during summer months, when it is so hot that many people spend hours sitting out on their porches, trying to keep cool. Why not design better cooling systems, the Rethinkers asked, or have students go to school earlier and return home for the hottest part of the day?

Then they turned to food—in particular, the use of oil to transport food. "The students saw the urgency of getting involved in school gardens after this exercise," recalls Burkes. They also recognized the need to increase the production of local food—a point they made to ARAMARK, the district's new food service provider. And in 2011, ARAMARK signed a memo of good faith to provide New Orleans schools with two locally harvested fruits per week.

The students, meanwhile, also began to examine their own habits and how they might reduce their own dependency on oil. They talked, for example, about using their cell phones less often and going to see friends more often, swapping clothes instead of buying new ones, and biking more instead of being driven to school. And then they signed pledges to put their words into action.

"We have to make learning personal for students to be engaged," says Burkes. "And I think the notion of emotional, social, and ecological intelligence really helps with that. Preparing lessons that we want to teach, preparing materials that we want to use—it's very helpful to remember that there is an emotional intelligence and a social intelligence that we want to reach for." It is the key, she adds, to making learning personal and relevant.

After reading "Beyond Whining," consider the Rethinkers' conclusion that we all bear ultimate responsibility for the Gulf oil spill and the resulting damage, due to our dependence on oil. If the Rethinkers are right, what are the implications for you? What are the implications for your students?

PART THREE

SHARED WATER: MOVING
BEYOND BOUNDARIES

Water Wars and Peace

Chapter
FIVE

Aaron Wolf, Mediator and Oregon State University Professor

This story reveals how moving beyond just the facts of an issue and acknowledging the emotions, needs, and experiences of others can lead to unexpected opportunities for cooperation. Developing empathy for other life forms (both human and non-human), one of the five practices of socially and emotionally engaged ecoliteracy, carries great transformative potential. As you read, imagine being the voice for the many non-human life forms that cannot speak their needs but that are, like us, dependent upon water for their survival.

W HEN PEOPLE COME TOGETHER for a common purpose—whether they are students working on a new class project or warring nations negotiating the use of shared water resources—there is always the potential for something Aaron Wolf refers to as a "sudden jolt." "People in education and mediation, they know the jolt," says Wolf, an Oregon State University professor of geography and a mediator of water disputes around the world. "They all recognize it." There is a distinct, transformative shift in the room when suddenly everybody sees, understands, or experiences things differently from the way they did before. Very often, he suggests, it is the "Aha!" moment for which teachers and mediators live.

But what exactly leads to such moments of transformation? Is it possible to design the conditions in a room, and engage with the people in it, so as to bring about these moments more often? And if so, what does one need to understand about the transformation process to work with it? These are the questions that have fascinated Wolf in his work as a scientist, an educator, and a mediator of water conflicts from the United States to the Middle East.

Throughout his journey, Wolf has recognized the essential role of emotional intelligence, social intelligence, and ecological intelligence—as well as spiritual intelligence. To be sure, it is an unconventional approach for a scientist, but one Wolf has found central to a deep understanding of human behavior regarding ecological issues. Indeed, he believes that to develop realistic solutions to some of today's great challenges, "you need an understanding of both sustainability and the core emotional values that cause people to make the decisions they make."

"I originally thought I would save the world through better technical models," Wolf adds with a smile. "But I kept getting drawn to the human dimension." Hired to map springs around Madison, Wisconsin, for example, he routinely turned in reports late because he spent so much time talking to landowners.

In time, he came to realize that "the scientific approach really focuses on understanding physical symptoms, pointing to better efficiencies, and developing better handles on cause and effect. It can then also point to the things that people ought to do. But if you don't take into account what people care about, what they are passionate about, why they make certain decisions, you only understand half the puzzle."

As a result of his own efforts to take a broader view, Wolf has flipped a widely held assumption about water wars on its head. He has also begun to change the discourse about what some have called the defining issue of the century. As a result, he has helped warring nations turn the need to share resources to a pathway to peace. And these efforts have helped him serve as a more effective facilitator—both in negotiations and in the classroom.

LIKE OIL, LIKE WATER?

For the past twenty-five years, many officials have predicted that nations will eventually go to war over water as they have over oil. Dr. Boutros Boutros Ghali, former Secretary General of the United Nations, first declared in 1985, "The next war in the Middle East will be fought over water, not politics."[1] In 1995, Ismail Serageldin, then Vice President for Environmentally and Socially Sustainable Development at the World Bank, announced, "The wars of the next century will be fought over water."[2] A few years later, Kofi Annan, another former United Nations Secretary General, said, "Fierce competition for fresh water may well become a source of conflict and wars in the future."[3] And more recently, United Nations Secretary General Ban Ki-moon proclaimed water scarcity to be a "potent fuel for wars and conflict."[4] The U.S. Central Intelligence Agency has similarly warned of "water wars."[5] Numerous books have invoked the same specter.[6] And documentaries such as *Blue Gold: World Water Wars* have warned that ordinary citizens may soon have to fight for water.[7]

These predictions, of course, do not come out of the blue. They emerge from growing concerns about trends that are exacerbating the scarcity of clean drinking water in developing countries—and threatening to bring an end to what journalist Charles Fishman has described as the developed world's "golden age of water."[8]

Today, one out of six people in the world lacks access to clean, safe drinking water. Many of them are children. In fact, if you picture an American elementary school of 500 students, and then imagine ten such schools all lined in a row, you would be looking at roughly the number of children who die every day from lack of access to adequate clean water, as Fishman reports in *The Big Thirst: The Secret Life and Turbulent Future of Water.*

And what of developed nations? For the past century, the majority of people in America and other developed countries have known water to be abundant, safe, and cheap. Indeed, the water we use—to drink, cook, bathe, clean our clothes, grow our gardens, and refresh ourselves and our children on hot days—has been so reliable that, Fishman observes, we don't even have words to describe the aquatic equivalent of "power failure." Like the air we breathe, we have learned to count on it without even thinking about it.

THE BIGGEST (HIDDEN) USE OF WATER

When we think about how we use water in the United States, we usually first think of the water we use at home: for drinking, cooking, washing—and flushing the toilet (which requires more water than the other three activities combined).[9] After that, we might think about irrigation systems on farms, lawns, and golf courses. But what we often miss are the hidden uses of water—even though one of them represents the biggest use of water in the nation and one of the largest worldwide.

"Water is the secret ingredient of our fuel-hungry society," wrote journalist Charles Fishman. In fact, our use of electricity consumes two-and-a-half times more water than the amount used in our bathrooms, kitchens, and yards.[10]

In the United States, more water is used in the production of electricity than anything else. An estimated 201 billion gallons of water a day were used to power thermoelectric plants in 2005, the last year for which these estimates are available from the United States Geological Survey.[11] Most of this water is used to cool the equipment.

Irrigation, or the watering of crops, accounts for the second largest use of water, at 128 billion gallons a day. And unlike water we use in our homes—90 percent of which ends up back in streams or groundwater supplies—about half of this water is lost to the natural process of evaporation or evapotranspiration (which includes the release of water through plant leaves[12]), or as a result of leaking pipes.[13]

Public water use—withdrawn by water departments from rivers, lakes, reservoirs, and wells for delivery to homes, businesses, and schools—represents the third largest use of water at 44 billion gallons a day.[14]

Other top uses include water used for industry (eighteen billion gallons a day), aquaculture or fish farming (nine billion gallons), mining and domestic uses through public and private supplies (four billion gallons each), and livestock (two billion gallons).[15]

But all the same, the reality is this golden age is coming to an end. "We are entering a new era of water scarcity—not just in traditionally dry or hard-pressed places such as the U.S. Southwest and the Middle East, but in places we think of as water-wealthy, such as Atlanta and Melbourne," Fishman writes. In developing countries, of course, the problems will get even worse.

CAUSES OF SCARCITY

Three of the biggest trends driving water scarcity are population growth, economic development, and climate change.[16] Global population has increased from about 1.6 billion in 1900 to 7 billion in 2011. By 2045, the United Nations predicts, it will reach 9 billion. And although every person on Earth needs water, we have only a finite supply—the same amount we have always had, endlessly recycled from the clouds, to waterways and oceans, to our homes and back again.

Moreover, there is no way around our limited global supply: We cannot make our own water, nor can we turn to water substitutes (although increasing efforts are underway to turn salt water into fresh water).

Every one of us requires a minimum of three liters of drinking water a day. One liter of water is also required to produce the food that supplies us with one calorie of food energy—or 2,000 liters (528 gallons) for each day's recommended calorie intake per person.[26] And that just scratches the surface of our water demand.

OCEAN WATER IN A GLASS?

Ocean water, which represents 97 percent of the Earth's water, has long tempted those who have wrestled with shortages of fresh water.

The earliest known efforts to remove salt from ocean water date back to 2000 B.C., according to ancient Sanskrit texts. Hippocrates also spoke of it, as did Sir Francis Bacon.[17] And in 1962, president John F. Kennedy declared, "If we could ever competitively—at a cheap rate—get fresh water from salt water, that would be in the long-range interest of humanity, and would really dwarf any other scientific accomplishment."[18]

Sixteenth century ocean voyagers are thought to have made the first serious efforts to desalinize water using simple devices on ships.[19] But the first large-scale desalination plants only began in the 1970s.

Today, an estimated 300 million people in 150 countries use water from the sea or brackish groundwater—about twice the number who did in the 1990s.[20] An estimated 13,080 desalination plants exist around the world and produce an average of twelve billion gallons of water a day.[21]

About half of these projects are located in the Middle East, although other regions have also begun to turn to desalination in recent years. For example, a $300 million facility has been

In fact, due to economic development, growth in water use has exceeded population growth by a rate of two to one during the past century, according to the United Nations.[27] Water is also required for the manufacture of almost all products, including those made out of metal, wood, paper, chemicals, gasoline, and oils.[28] And since the turn of the twenty-first century, economic development has surged—most notably in China and India, two of the world's most populous nations.

Climate change is also driving scarcity by disrupting nature's water cycle in numerous ways—causing wet regions to get wetter, dry regions to get drier, droughts to last longer, flooding to be more intense, and snow to melt faster. Climate change leads to drier soil, which slows the replenishment of groundwater. And perhaps most dramatically, it accelerates the melting of glaciers, affecting the water supplies on which millions of people depend. "Millions if not billions of people depend directly or indirectly on these natural water storage facilities for drinking water, agriculture, industry, and power generation during key parts of the year," according to Achim Steiner, Executive Director of the United Nations Environment Programme.[29]

As a result of these and other factors, the United Nations forecasts that by 2025, two-thirds of the world's population could be living in water-stressed conditions—where they might find it difficult or impossible to get adequate water to meet the needs of agriculture, industry, and households.[30]

proposed near San Diego, California. A $2.9 billion facility is planned near Melbourne, Australia.[22] And in Tianjin, China, a $5 billion plant is projected to be China's largest. "As it did with solar panels and wind turbines," Michael Wines reported in the *New York Times*, "the government has set its mind on becoming a force in yet another budding environment-related industry: supplying the world with fresh water."[23]

With projections that one billion urbanites could face water shortages by 2050 due largely to urban growth and climate change, a rush to build systems capable of converting massive amounts of salt water into drinking water makes sense.[24] Numerous technological advances also have helped bring down the costs.

But there are several associated ecological concerns. Most notably, the process of drawing enormous amounts of seawater into desalination plants destroys the habitat for countless fish, plankton, seaweed, and other marine life.

In addition, every fifteen to fifty gallons of drinking water produced from saltwater leaves thirty-five to fifty gallons of concentrated brine, according to Alex Prud'homme, author of *The Ripple Effect: The Fate of Freshwater in the Twenty-first Century*. And that raises questions about what to do with all the brine that is left behind.[25]

CHALLENGING THE ORTHODOXY

Given these realities—and the fact that two out of five people in the world live in places where rivers and lakes cross national boundaries—it is easy to imagine nations going to war over water. But have they ever done so during periods of water scarcity in the past? This is the iconoclastic question that Wolf asked—in part because his scientific training taught him to search for evidence, but also because, based on things he had seen in his childhood, he suspected that the experts might have it wrong.

Wolf grew up (with his sister, the feminist writer Naomi Wolf) in San Francisco during the politically charged 1960s and 1970s. His father, Leonard Wolf, was a writer and professor; his mother, Deborah Goleman Wolf, was an anthropologist.[31] And both instilled him with a strong sense of ethics. "The idea of our obligation to right wrongs was clearly the language of the day," he says.

The years 1976 and 1977 saw the worst drought in California's history, a time when skateboarders turned abandoned pools into makeshift skate parks. Wolf became fascinated with water during these years, aware of its potential to lead both to conflict and cooperation. But he experienced the greatest influence on his future work during the few years his family spent living in Israel.

The Middle East is the most severely water-stressed region in the world—one where unfriendly, even warring neighbors must work out ways to share this scarce resource.[32] The people of Turkey, Syria, and Iraq, for example, all depend on water from the Euphrates River. The people of Egypt and the Sudan rely on water from the Nile. And the people of Israel, the West Bank, Lebanon, Syria, and Jordan must share water from the Jordan, perhaps the source of greatest contention.[33] "For a biblical stream whose name evokes divine tranquility," Don Belt recently observed in *National Geographic*, "the Jordan River is nobody's idea of peace on Earth."[34]

The enemies there are intense enemies, Wolf says. "Yet while living in Israel, I saw so much cooperation that no one had been talking about. I thought: If there is so much implicit cooperation going on in the Middle East around water, if it isn't true that people are going to war over water there, where is it true?"

UNCOVERING THE FACTS

Working with a team of researchers from Oregon State University, Wolf spent three years studying all known conflicts between nations in which water was a driving factor. His findings led him—hesitantly at first and, eventually, more stridently—to challenge the prevailing opinion about water wars, until one day he declared during a lecture in Norway, "There was never a war over water in all human history."

A man in the back of the room, whom Wolf describes as a ruffled-looking historian, raised his hand and asked, "When you say 'never,' just how far back are you going?" It was his way of advising Wolf that there had been a water war between two Mesopotamian city-states, Lagash and Umma, some 4,500 years ago.[35]

Water is the most recyclable of all resources. The water that appeared on Earth some 4.5 billion years ago is the same water that flows in our oceans, rivers, and streams today. And, of course, it is the same water we use to drink, cook our food, water our plants, and bathe our children. Every drop has a staggeringly long history.

Your most recent glass of water, for example, could have flowed through a dinosaur, mammoth, or saber tooth tiger. It could have quenched the thirst of the builders of Stonehenge or the Taj Mahal. It could have refreshed the first Olympians or the last inhabitants of Easter Island. It could have touched the skin of Jesus, Muhammad, or the Buddha. And it most certainly floated aloft in clouds and moved with fish through countless streams.

This history is a testament to what may be the most remarkable qualities of water: It cannot be used up, and it cannot be destroyed. It is resilient beyond compare. And given the many ways in which we humans pollute water—including the runoff of chemicals from industrial farming, slurry from coal production, and waste dumped in streams and rivers—this is good news, indeed.

Water is also cleaned naturally through evaporation, crystallization, and aeration (which occurs, for example, when a fast-moving stream crashes repeatedly into a rock and becomes an air-borne spray).[36]

More recently, people have developed techniques to disinfect water more quickly than nature—albeit on a vastly smaller scale. Perhaps the most simple and popular method, used by an estimated five million people worldwide, is solar water disinfection, or SODIS.[37] Endorsed by the World Health Organization, it uses UV radiation and the heat of the sun to inactivate waterborne microbes that cause diarrhea and other waterborne illness.

To try this in your home or school:

1. Remove any solids by filtering the water or allowing it to settle.

2. Fill the water in one- or two-liter clear plastic bottles.

3. Shake well to aerate.

4. Expose the bottles to sunlight for five hours (or up to two days if it is cloudy).[38]

But that one long-ago war over water appears to be the exception to the rule. As the Woodrow Wilson International Center for Scholars concluded, "Exhaustive research by Aaron Wolf of Oregon State University has firmly established that international violent conflict is rarely—if ever—caused by, or focused on, water resources."[39] What Wolf found instead was evidence of approximately 3,600 water-related agreements.

More specifically, of 1,831 conflicts between nations over water during the period from 1946 to 1999, only 5 percent (93) led to physical hostilities. An estimated 23 percent (414) involved verbal hostilities. And the overwhelming majority—67 percent (1,228)—resulted in peaceful resolutions.[40] Instances of cooperation,

Wolf points out, outnumbered conflicts by more than two to one.[41] (Other researchers have since affirmed Wolf's findings.[42])

What the facts about water conflicts reveal, says Wolf, is this: "People will go to war over oil, over gold, over diamonds. But water is different. It is really hard to think about depriving your enemy of water, even as you are depriving your enemy of other resources." During the first Gulf War, for example, Turkey allowed the United States and its allies to launch air attacks on Iraq from its territory, but it refused the allies' request to shut off the flow of the Euphrates River. "In their mind, you could use air bases to launch bombing raids. But shutting off water was too horrific to consider."

Still, even if nations do not go to war over water, Wolf explains, all the other problems that arise from the need of a growing number of people to share a limited supply of water remain. "People suffer. There is ecosystem degradation. There are high levels of conflict, tensions, and violence."

And yet, Wolf's point is that is exactly where the opportunity for transformation is created. "Even when people are not shooting at each other, they still need to come together to figure out how to manage the water that's there," he observes. And it is out of that need to come together that water can serve as an elixir of peace.

THE SPACE BETWEEN WAR AND PEACE

What sparked Wolf's quest to work the space between war and peace began with an observation: While serving as a mediator of water conflicts, he would show those engaged in the conflict two maps—one of countries defined by conventional political boundaries, and another of the same territory with boundaries defined only by water and land. From this experience he often sensed that people suddenly began to see their world differently. No longer were they thinking only about their own agendas, but they were recognizing and understanding the needs of others.[43] In the Middle East, for example, "they [momentarily] forget about distinctions between Arabs and Israelis, because the only thing on the map is what unites them."

Why exactly did this exercise prove to be such a powerful catalyst of transformation? When Wolf studied the literature of conflict resolution, he was unable to find a persuasive answer. Then one day, he happened to be talking with World Bank water advisor and friend, Vahid Alavian,[44] who observed that the images seemed to reflect a metaphor for spiritual transformation.

> We usually start with our own map of the world. We go through life with our own needs, wants, and expectations, which form the boundaries of this map as we individually set them. If we learn to remove or at least adjust those boundaries, a new map of the world can be drawn based on relationship with the community around us. The key to mediation in this context is to transition from one map to the other with newly defined needs, wants, and expectations for the entire

community and its prosperity. This transition requires examining and introducing elements beyond the current practice to include the moral and spiritual dimensions.[45]

The possibility of studying spirituality for insight into transformation hooked Wolf in that moment for several reasons—not the least of which, he quips, was that he had the tenure that allowed him to study whatever he chose without threatening his career. More significantly, his friend's observation helped him realize that spiritual traditions were perhaps the most natural places to look for an understanding of the process of transformation. After all, he explains, every spiritual tradition in the world is devoted to guiding individuals to think less about their own immediate wants and desires and more about others.[46]

So Wolf spent the next year traveling and studying. He began in Israel, where he studied Judaism and the Kabbalah, and later traveled to Thailand, where he learned from Phra Paisan Visalo, a Buddhist monk who mediated conflicts between villagers and the timber industry. He also read widely to broaden his understanding of other world traditions.

But although the subject was a mystical one, and spirituality had always been important to him in his personal life, Wolf's goal in this quest was strictly pragmatic. "It wasn't a study of mysticism for its own sake," he explains. "What I was after was, What is useful? What can we apply to the conflict-resolution world? What can we learn from mystical experience that we can bring into a room of angry people?"

One thing that became clear to him was that the dominant Western perspective on what influences human behavior is too narrow. "There are so many things we talk about with certainty—ideas we think we can quantify, such as intelligence, or the economy, or affection. As soon as there is a number attached, we think we're okay. But the limitation of that shows up regularly. Especially when you try to describe the conflict resolution process, rationality falls apart immediately because people also make decisions that are counter to rational values."

People in many parts of the world understand this, Wolf observes. But people in the West often forget it because our intellectual history severed the worlds of spirit and reason.[47] In practice, he says, that means Westerners sometimes think that people will agree to a certain action only when it is in their economic, environmental, or strategic interest to do so. But he insists that those motives alone fall short of a complete model for understanding human behavior regarding water or any other natural resource.

A NEW (OLD) MODEL

According to Wolf, a truer—indeed, a universal—model for understanding human behavior is what is known as the "Four Worlds" (or what historian Huston Smith called "levels of reality"[48]). These levels of reality, or lenses through which we experience ourselves in relation to others and the world, include the physical, emotional, knowing, and spiritual. Each is important. Each is valid. And all can be true at the same time. "Psychologists will recognize [Abraham] Maslow's hierarchy of needs[49] in the Four Worlds," Wolf adds. "But those familiar with the mystical traditions around the globe will find much more ancient roots."

In a 2008 article, Wolf offered this illustration of the Four Worlds frame in practice:

> One intuitive example might be seen through a glass of water, which exists most recognizably on a physical plane. However, if one is thirsty or the water is particularly satisfying, one's experience of water can be transformed into an emotional response. One can also intellectualize the water and thus can consider its components and interaction of the water with our body to provide and maintain sustenance. Finally, one might say a blessing over the water, lifting its "profane" covering, and it now becomes a source of spiritual nourishment.[50]

These different ways of seeing the world help reveal that water, like any natural resource, can mean different things to different people. And in the hands of a skilled facilitator, this understanding can move people away from considering only their own interests and help them recognize the needs of all who depend on the same scarce resources.[51]

PUTTING FOUR WORLDS TO WORK

Today, Wolf applies the physical, emotional, knowing, and spiritual model in all aspects of his life, from the classroom to conflict negotiations. He has frequently used it, for example, to help resolve conflicts between Arabs and Israelis. "For the Palestinians, Gaza is one of the worst places for water in the world. There's not enough, and what's there is close to being poisoned; it's laden with pesticides and saltwater. When they talk about water, they're talking about survival on the physical level," Wolf says. In contrast, water issues are intellectual for the Israelis, who focus on such questions as how much water can they divert from agriculture to industry, and what would the impact be? "They're all using the same word," he adds. "But if you are dying of thirst, you can't treat the discussion like an intellectual exercise."

In one particular conflict, Wolf used the physical, emotional, knowing, and spiritual frame to help Israelis and Palestinians prioritize their respective needs for water. "It was not so easy," he recalls. "Farmers needed water for farms, environmentalists for the environment. For everybody, their need was most important." So Wolf acknowledged that the first priority uniting everyone in the room was water for drinking and spiritual purposes. "It was one of those seminal moments in the room. To recognize the spiritual significance in the same breath as drinking water was an instant symbol of respect toward each other's traditions and also a way to express explicitly how deeply felt the water issues are." In fact, once those priorities were clarified, the parties in conflict were able to work out many of their other issues—agreeing, for example, that water for subsistence farming should be given a higher priority than water for a semiconductor plant.

Wolf has also applied the physical, emotional, knowing, and spiritual perspective in a training manual he produced for UNESCO, *Sharing Water, Sharing Benefits: Working Towards Effective Transboundary Water Resources Management,*[52] and in a textbook, *Managing and Transforming Water Conflicts.*[53] But neither book is explicit about its inspiration. "If the Four Worlds perspective is where science leads me," Wolf says, "I have a scientific obligation to go there. If I stop because of some preconception or bias, that's not good science."

Still, he is practical enough to recognize that some people will just shut down at the mere mention of spirituality. "Face-to-face, I'm more comfortable being open about this," Wolf says, "because with a safe container, many people are cognizant about what we lose from not being explicit about our spiritual lives." In fact, when he speaks about the role of spirituality in his own life, people often look around and whisper, "Yeah, me too."

But whether or not he opens up about what he has learned from spiritual traditions when pursuing his work in science, mediation, or the classroom, the approach still works. "That's what's beautiful about this stuff," he adds. "These truths are so central and universal, they almost don't have to be named."

This story describes the profound shift in perception that can occur when one looks beyond political boundaries and considers the universal human connections to local bodies of water. Consider where you live. With whom do you share the water? What territory constitutes your watershed? Who manages it? Does the watershed cross political boundaries (such as cities or counties)? What personal and emotional connections do you have to bodies of water in your area? Do any local groups or events express an emotional or spiritual connection to water?

From Restoration to Resilience

Chapter
SIX

Students and Teachers Restoring A Watershed, Northern California

This story demonstrates how educators can help young students cultivate empathy for other forms of life (even the less-than-cute ones) and transform that empathy into action that deepens their learning and sense of community. As you read, notice how, with patience and perseverance, educators and students develop empathy, community, and an understanding of the ways nature sustains life.

Four women were visiting a ranch in Sonoma County, California, when a herd of dairy cattle began to get curious. At a distance, the animals looked as peaceful as any cows you'd see munching grass in a pasture as you drove by. But as the herd moved closer and still closer, the women began to feel mildly unsettled—and then, at the sight of one distinctly unhappy-looking bull, very definitely nervous.

There is no such thing as a totally safe bull, as Temple Grandin writes, and dairy bulls are the most dangerous.[1] So when this one approached with its head lowered, eyes protruding, and hooves pawing the dusty ground, the women—who were not ranchers, but ecosystem restoration practitioners—clearly felt the danger signs.

"I don't like the look of him," announced Emily Allen, a former AmeriCorps tutor who now works with K–12 student groups on restoration projects.

Instantly, the women scattered: Allen and her colleague Stephanie Nelson got behind a fence that safeguards a creek, while Vanessa Wyant and Laurette Rogers chose the nearest possible option and hopped into Wyant's small white pickup truck.

With this distance between them, the bull calmed down. But some 50 cows were not so easily appeased. They circled the truck and firmly planted themselves, as only cows can, blocking it from the front and back. Worried that they might be mistaking it for a food truck and begin to butt it if they were not soon fed, Wyant began to slowly drive away. But the cows slowly and resolutely followed. Wyant then drove a little faster. The cows moved a little faster. And on it went until both Wyant and the cows moved more than a football field away. Then, suddenly, Wyant made a U-turn and sped back to where the two other women watched from behind the fence. Finally, the cattle gave up the chase, and the women got back to business: monitoring and maintaining the creek restoration work young schoolchildren had carried out under the auspices of the nonprofit project, Students and Teachers Restoring a Watershed (STRAW).[2]

Ecosystem restoration work is, of course, not always this exciting. It is usually a slow, quiet process that requires great patience. Results can take months, even years, to appear—nearly a lifetime from a child's perspective. The work is deceptively complex; the science of ecosystem restoration is still in its infancy; and the collaboration required can bring together the most unlikely of partners. In other words, it can seem unlikely work in which one can successfully engage young students.

Nevertheless, STRAW has a twenty-year track record of success. It has engaged more than 230,000 students in restoring nearly twenty-one miles of habitat in California's Marin and Sonoma counties. It offers hands-on engagement that widens empathy, emphasizes leadership and working in community, and cultivates an understanding of ecosystems.

And in 2011, it embraced an urgent new challenge at the cutting edge of ecological restoration: giving students an opportunity not only to restore ecological functions that have been lost or damaged but to build in resiliency that reduces vulnerability to climate change.

"You can get kids to the point of despair, and then you have to face, what are you going to do?" says former teacher Laurette Rogers, who is the director of STRAW (a project founded under the auspices of the Center for Ecoliteracy and The Bay Institute and now part of PRBO Conservation Science). "I've talked to other teachers who get pulled into environmental education the same way. The kids get so depressed that the teacher feels compelled to do something."

But the ability to move beyond despair and engage in constructive, hopeful action in the face of today's significant environmental challenges rests, Rogers and others have found, on integrating emotional, social, and ecological intelligence.

"You need the emotional and social scaffolding for learning," says former principal Sandy Neumann, who now serves as STRAW's professional development manager.

Moreover, this form of engagement can be generally positive—for education and the environment. "I think people want to be involved in life-giving things: the environment and each other," Neumann adds. "And the process that we see in engaging teachers and students in problem-solving is exactly the same process that we see in restoring nature. It's about respecting the process that is already there, getting the things that are in the way out of the way, and letting things happen."

CULTIVATING COMMUNITY

When Neumann became the principal of Brookside School in San Anselmo, California, she had a dream of creating a learning community that emphasized active learning and shared decision-making, and that was intimately connected to the larger environment in which it resided.

"I was raised in northwest Michigan. I'd lived in the woods. And I was influenced by David Orr's writings[3] that all education is environmental education," Neumann said from her home, where, just a few moments earlier, a wild turkey had crossed her porch. "I knew that if our kids were going to understand themselves in the world, they would need to understand their connection to the environment."

Her teachers, however, were not so sure. "They didn't want to hear about it," she recalled. But Neumann, sympathetic to the often outsized demands facing teachers, had no intention of imposing her vision. Rather, she sought to patiently and gently attract them to her vision. ("If you value process, you have to value its rhythm and time," she said.)

So she allowed teachers to voice their opposition while she spoke about her vision and modeled the healthy, vibrant school community she wanted. She fostered opportunities for teachers to collaborate. She rotated leadership and taught that everyone had something essential to contribute. She applied for foundation grants to support teachers' efforts, even before they were on board. She made sure there was always plenty of food at meetings. And she consciously facilitated meetings that brought out the best in people and promoted conflict resolution. "There are always conflicts, and the whole point of a system in a sense is the working through of conflicts so the energy is there for creativity."

Toward the end of her first year as principal, a few teachers began to find unique ways to connect their students to the environment. Teacher Joan Owen, for example, took her students for a walk up a hill behind the school, where she had each student pick out a 3-by-3-foot piece of land. Then, every week, Owen and her students returned to the same spot to study the small patch in depth. "They would look and try to figure out what all the insects and plants were. They wrote poems and stories. They took their parents," recalled Neumann. "In fact, they got so deeply attached that when one kid's dad died, the student asked his whole family to come to his land to honor him."

A STUDENT'S QUESTION

While a fourth-grade teacher at Brookside, Laurette Rogers made her first small effort to connect her students with the larger environment by showing them a *National Geographic* film about rainforests and endangered species.[4] "It was filled with haunting music and pictures of chainsaws," recalls former Brookside student Aaron Mihaly.

When the film ended, the eight- and nine-year-olds sadly discussed what they had learned, slowly sinking lower and lower in their seats. Then one student, John

Elliott, raised his hand and said, "Mrs. Rogers, what can we do to help endangered species?"

Rogers says, "I looked at him and knew I couldn't give a pat answer and say 'Let's write letters' or 'Let's do a study.'" Indeed, she was so excited that slept uneasily for two weeks.

Finally, she reached out to the California State Adopt-A-Species Program for help with devising a project for her students.[5] She then presented three options to the students for a vote: they could adopt trout, salmon, or the California freshwater shrimp.

None of the options represented the soft and cuddly creatures young children often fall for, but the students had to choose one so they settled on the shrimp. Her theory was that if the students found out all about the shrimp—or any particular species for that matter—they would fall in love. So, over time, the students learned that the shrimp are beautiful, almost transparent creatures that have lived in local creeks since the time of the dinosaurs. The males grow up to 1.5 inches long, and the females grow up to 2.5 inches long. They are the creeks' garbage collectors, feeding on dead and decaying plant material. They are also terrible swimmers that must cling to riparian plant roots so as not to be washed away.

Rogers gave her students original scientific papers to read and made each fourth-grader responsible for deciphering and accurately reporting the most important information from one to two pages. Students analyzed the data for each of the fifteen creeks in which the shrimp live. And they created shrimp drawings during art time and shrimp poems during language arts.

They also learned that the shrimp are threatened primarily because of habitat destruction around the streams where they live. Dairy, beef, and sheep ranches are the agricultural mainstays of west Marin and Sonoma Counties. In former years, agricultural agents used to advise dairy farmers to build their pastures near creeks to provide water for their stock. Now, the students discovered, the shrimp habitats were pressured by the damming of creeks, petroleum and chemical runoff, manure in the water, and sedimentation from soil erosion caused by animals trampling the creek banks and grazing the foliage that could otherwise stabilize the soil. Off-road vehicles, trash dumps, and damage incurred by potato farmers were also negatively impacting the stream environment.

And shrimp, the students learned, were not the only species affected. These creatures were merely one strand of a web that included trees, grasses, aquatic insects, songbirds, creeks, estuaries, and the San Francisco Bay itself. The students began to understand the "shrimp problem" as a watershed problem. They also saw how the story of "their" shrimp was repeated over and over again for other endangered species.

Finally, after nearly six months of study, the students saw a shrimp for themselves. And when they did, Rogers recalls, they exclaimed "Ahhh!" as if they'd seen a movie star. They had, indeed, fallen in love.

Frogs, mussels, and the 126,000 other animal species that live in freshwater lakes, swamps, and rivers are disappearing at a rate four to six times as fast as animals on land or at sea. In the United States, for example, approximately half the 573 animals on the threatened and endangered list are freshwater species, according to *National Geographic*.[6]

"That's because freshwater ecosystems are so closely linked to human activity," Douglas H. Chadwick wrote in the magazine's special issue on water. "Industry and agriculture are concentrated alongside flowing waters, and sooner or later the residue of virtually everything we do winds up running down the nearest creek—if we haven't dried up the creek first."[7]

Worldwide, the growth of dams is having an even greater impact. In 1950, there were 5,270 large dams worldwide, according to the World Resources Institute.[8] Today, more than 45,000 large dams and many more small dams interfere with freshwater ecosystems by blocking pathways used by migrating fish, reducing and rearranging the patterns of flowing water that have shaped aquatic life for millennia, and changing water quality.[9]

But numerous organizations such as International Rivers, the Nature Conservancy, and Conservation International are working to protect freshwater ecosystems. Students and teachers are also showing what they can do through projects such as Students and Teachers Restoring A Watershed (STRAW).

THE WATERSHED CONNECTION

Watersheds provide one of the clearest examples of the interconnectedness between human activities and ecosystems. Healthy watersheds provide clean drinking water, habitat for fish and wildlife, and reduced vulnerability to severe impacts from invasive species, climate change, and future adaptations in land use. But nationwide, the quality of our freshwater streams is still in decline, according to the Environmental Protection Agency. For example, a recent EPA survey of the nation's wadeable streams found 42 percent in poor biological condition and 25 percent in fair condition. Nearly 40 percent of North America's freshwater fish, 700 species in total, are imperiled.[10]

One of the many causes of watershed degradation that has become increasingly important in recent decades is fertilizer and other runoff from industrial agriculture and ranches—and the creek Rogers and her students chose as their focus was a case in point. Stemple Creek is a sixteen-mile-long creek that flows from the hills of Petaluma, through about ten miles of northern California cattle ranches, before passing into Estero de San Antonio, a stream that, in turn, empties into Bodega Bay, which is part of the Farallones National Marine Sanctuary on the Pacific Ocean.

It is considered nearly impossible to restore a dead zone. But it is possible to stop their proliferation. Two good ways involve transitioning off fossil fuels and finding ways to keep fertilizer on the farm.

In the Gulf of Mexico, off the Louisiana coast, lies an area about the size of New Jersey, where all the fish have fled, and clams and other marine life that cannot swim have died. This region of more than 8,000 square miles is the largest "dead zone" in the United States and one of 405 dead zones worldwide.[11]

"Dead zones were once rare. Now they are commonplace," according to Professor Robert Diaz of the Virginia Institute of Marine Science, College of William and Mary.[12] Since the 1960s, when forty-nine dead zones had been identified, the number of zones that offer too little oxygen for most marine life to survive has approximately doubled every decade.

All told, the combined area of global dead zones (95,000 square miles) is relatively small in comparison to the 71 percent of the Earth's surface covered by oceans.[13] But more than 83,000 tons of fish and other ocean life either die in or flee from the Chesapeake Bay dead zone each year. More than 235,000 tons of life are lost in the Gulf of Mexico dead zone. And in one year alone, a dead zone off New York and New Jersey caused fisheries to lose an estimated $500 million in income.[14] Dead zones, in short, are thought to be the most significant threat to marine ecosystems today.[15]

The primary culprit is industrial agriculture—or, more specifically, the excess fertilizer that runs off farmland. Fertilizer is rich in nutrients (primarily nitrogen and phosphorus). But when

The Estero de San Antonio is considered one of California's most significant habitat areas because of its rich mosaic of densely wooded riparian ravines, salt grass areas, mudflats, eelgrass beds, and small freshwater ponds. The EPA, however, has also declared it an impaired body of water due to excessive nutrients and sediment from nearby pasturelands and feedlots.[21]

After settling on Stemple Creek as the focus of their restoration plans, Rogers' students made presentations to a meeting of the local Resource Conservation District and other community members. They also wanted to meet with local rancher Paul Martin, who shared their concern about erosion.

Martin was initially a little skeptical about allowing a group of fourth-graders onto his property—and decidedly leery of environmentalists telling him how to run his business. But then he learned that Rogers had told her students to imagine what it would be like if someone came into their bedroom and said, "From now on, you can't get anything out of your closet—none of your toys, clothes, or anything." It would be similarly unfair, she said, if they or anyone else told the rancher what to do with his land. "After I heard that story, I knew it would be all right, and we started working together," Martin has said.

The class completed its first project on Martin's ranch in 1993, planting willows and oak trees along the creek banks. "In our area, you get more bang for the buck with willows than anything else," says Rogers. "Students can see results. In four months, the sprigs they plant will have branches three to four feet long. In two years, they'll look like little trees. They stabilize the soil. They provide shade fertilizer is spread on soil, only 10 to 15 percent of the nitrogen ends up in plants, while the rest washes into groundwater, streams, lakes, and eventually the ocean.[16] In the ocean, it has the unintended consequence of fertilizing massive algae blooms that block sunlight and starve the water of oxygen as the algae die, sink to the bottom of the ocean, and decompose.[17] For the forms of marine life that live there, two options exist: those that can, flee; those that can't, die.

Fertilizer use has also played a key role in our crossing the safe boundary level on two of the nine life support systems considered essential for human survival: It has accelerated biodiversity loss and interfered with the nitrogen cycle.[18] (The nitrogen cycle is a set of processes by which nitrogen is circulated between the atmosphere and organisms.[19])

The burning of fossil fuels, especially gas and diesel, also contributes to the formation of dead zones, as this process sends nitrogen oxide into the atmosphere—until it rains, that is, and the nitrogen falls into the ocean.[20]

It is considered nearly impossible to restore a dead zone. But it is possible to stop their proliferation. Two good ways involve transitioning off fossil fuels and finding ways to keep fertilizer on the farm.

to cool the water and reduce evaporation. Birds nest in them and bring in seeds of other trees like alders and oaks."

Every year since that first one, students have returned to Stemple Creek to work. The first plantings are now a tall, dense growth that blocks sight of the creek. The state bird, the California valley quail, has returned to the watershed. Songbirds are nesting in the trees. And, to everyone's surprise, the freshwater shrimp—which were not expected to reestablish themselves for decades—have migrated downstream and are clinging to the roots of the willows the students planted.

NEW DIRECTIONS IN RESTORATION

In the two decades since STRAW was founded, restoration has become a big business worldwide, as Margaret Palmer, director of the Chesapeake Biological Laboratory and Professor of the University of Maryland, has observed.[22] State and municipal governments frequently invest in a variety of restoration projects, from the removal of disturbances that interfere with natural stream functions, to the stabilization of streambeds and the installation of storm water management facilities.[23]

Until recently, however, it has been difficult to answer one important question: Are restorations good not only for biodiversity, as *Scientific American* recently put it, but for humanity?[24] José Benayas, a biologist at the University of Alcalá near Madrid, Spain, studied nearly ninety projects worldwide to determine how restoring ecosystems impacts and benefits the wild biological ecosystems as well as humans' biological systems, or "ecosystem services," such as clean water and air and recreational benefits. His findings, according to a study published in *Science*, indicated that restoration increased biodiversity by 44 percent and increased ecosystem services by 25 percent.[25]

THE NEW FACE OF WATER POLLUTION

Picture a glass of polluted water. In your mind's eye, chances are you see a glass of dirty brown water with a small mound of contaminants settled at the bottom of the glass. But in recent years, there have been growing concerns about the kind of pollutants you can't see, smell, or taste.

Perhaps the most notorious invisible pollutant is perchlorate, an ingredient in rocket fuel that finds its way into drinking water, largely as the result of improper disposal at rocket test sites, military bases, and chemical plants. Perchlorate in drinking water has been linked to thyroid and developmental problems, especially in newborns, young children, and pregnant women.[26] And more than sixteen million Americans have been exposed to unsafe levels of perchlorate, according to the Environmental Protection Agency.

Perchlorate, moreover, is only one of hundreds of chemicals found in Americans' drinking water between 2004 and 2009, according to an Environmental Working Group analysis of twenty million drinking water quality tests. Water suppliers also detected the following:

- Two hundred four industrial chemicals, including arsenic, lead, and dry cleaning solvents
- Ninety-seven agricultural pollutants, such as pesticides and fertilizers
- Eighty-six contaminants linked to sprawl and urban pollution, including mercury, benzene, lithium, and other pharmaceuticals[27]

The Safe Drinking Water Act regulates only ninety-one of the more than 60,000 contaminants in use today.[28] No chemicals have been added to that list since 2000. In 2011, however, the EPA announced that it intends to set a limit on at least one: perchlorate.[29]

Designing restoration projects with an eye toward preparing for climate change has been another equally important recent development. Restoration projects have traditionally benefitted climate change trends, as restored landscapes can help boost carbon storage capabilities that offset greenhouse gas emissions. But restoration specialists are now exploring how restoration projects can be designed to reduce vulnerability to climate change by making ecosystems more resilient.

"People throw up their hands and say, 'What can we do about climate change?'" says Tom Gardali, an ecologist with PRBO Conservation Science. "I would say restoration is one of the biggest things we can do."

STRAW launched its first set of restoration projects specifically designed to build resiliency in the face of climate change beginning in the fall of 2011.

"All the climate modeling, all the research shows there are going to be changes: birds will come at different times, temperatures will be different," says John Parodi, STRAW's restoration manager. "The way wildlife—and ultimately we—will be able to deal with it is by being able to adapt to these changes."

In other words, forward-looking restoration designs offer the opportunity to plan for drought- and fire-resistant foliage, and broaden the seasons in which animals can seek shelter and forage. That way, if the animals come sooner or later than usual because of temperature changes, there will be something for them in bloom.

THE ULTIMATE GOAL

Whether STRAW's restoration is focused on looking backward or forward in time, however, its primary goal always comes back to good education.

"From a conservation science perspective, I think the real long-term solutions aren't tweaking restoration design or buying land or managing land differently," says Gardali. "The real long-term solutions are when the populace can understand the consequences of environmental change and develop solutions-oriented thinkers. And I think STRAW does both of those things."

In carrying out STRAW's ecological restoration projects on the grounds of social and emotional intelligence, says Neumann, "modeling is the most important thing. We don't really talk about it. We just always do it."

STRAW works with teachers and students to provide scientific, educational, and technical resources to prepare them for hands-on watershed studies. Then, between November and March, they guide the students in carrying out the projects—usually, in the presence of parents and other community members.

"Sometimes parents joke that we probably have to come out here and redo it all tomorrow," says restoration manager Emily Allen. That's not the case, however, she says. "Once in a while, we might have to come out to fix a plant that wasn't planted right at the end of the day and we didn't want them to feel bad. But it's the students doing it."

Rogers says it's not about preparing students to be leaders of the future. "It's about seeing students as competent now."

It is equally, says Neumann, about supporting educators in times of growing ecological challenges. "One of our goals is to provide experiences that assist educators to find an inner path that will lead to deeper personal values and beliefs of interconnectedness and the value of biodiversity in all life," she says.

"The restorations and long-term results connect us with complexities at many levels and teach us that we can create the conditions for the Earth to heal herself."

Straw has worked at the slow process of restoring habitat for twenty years. Now they are facing the impacts of climate change on local habitats. Rather than ignoring or denying the implications of climate change, they are designing restoration strategies that build in safeguards for animal and plant species that are especially vulnerable. Think about the ways that their solutions-based approach to climate change requires integrating social, emotional, and ecological intelligence.

PART FOUR

NOURISHING COMMUNITIES WITH FOOD

Changing a Food System, One Seed at a Time

La Semilla Food Center, Anthony, New Mexico

This is a story about a group of young adults in New Mexico who set out to change the way people in their community nourish themselves. As in most of rural America, the land surrounding the community and the eating habits of its inhabitants were altered when small family farms were replaced by industrial agriculture. As you read, pay attention to the many ways that practices based on breakthrough scientific discoveries about how to grow food have resulted in unintended consequences, and consider the significance of developing the ability and discipline to anticipate unintended consequences (one of the five practices of socially and emotionally engaged ecoliteracy).

IT TAKES A CERTAIN KIND OF PERSON—or, in this case, group of friends—to look at fourteen acres of dry, dusty desert in one of the poorest regions in the country and envision a way for a community to take the food system into its own hands. Meet the three founders of La Semilla Food Center in Anthony, New Mexico, a small rural community near the Mexican border.

Cristina Dominguez-Eshelman is a soft-spoken thirty-four-year-old who didn't realize how important food was to her until she moved away from her family—or how much she loved growing things until she got her hands in some soil.

Aaron Sharratt is a gentle thirty-two-year-old who, while traveling as an undergraduate in Mexico, became hooked by a deep interest in how a landscape can impact what we eat, what jobs we hold, and even the relationships we have with our families and communities.

Rebecca Wiggins-Reinhard, thirty-one, is a spirited social activist who grew up completely uninterested in her family's farm—until she discovered that social justice issues were tied to every aspect of the food system.

89

The three came together several years ago around a modest project focused on engaging young people and their families in creating community gardens in Anthony and nearby towns. In the process, they say, they recognized the potential to inspire changes in eating habits, build awareness of food systems, and unleash the leadership abilities of young people—even in the face of some rather extreme everyday challenges.

"When you begin to think about the issues facing people here—oh, my gosh, the obesity problems, the public health concerns, the security, the border, the border patrol," says Sharratt. "And yet, there is so much energy, so much potential."

But how does one transform a sense of possibility into systemic change in food justice, health, and economics? "We thought if we're going to make an impact, if we're thinking about a kind of systems change at a big level, there would need to be people who are focused on driving those efforts forward," says Dominguez-Eshelman.

So the three friends left their jobs and created La Semilla Food Center, a nonprofit dedicated to bringing food justice and economic opportunity to the people of the Paso del Norte region of southern New Mexico.[1] The Center is establishing a farm, offering an education program on how to grow and cook nutritious food that is native to the region, and convening a Youth Food Policy Council that teaches young people how to influence the local food system.[2] Their goal is to reshape what they identify as the "foodshed" that stretches from El Paso, Texas, to Las Cruces, New Mexico.

The challenges they face are enormous. The old well on their land, once used to irrigate cotton fields, is broken. Sections of the wire fence around it were stolen. The Rio Grande is now no more than a big dry riverbed for several months a year. According to the New Mexico Environment Department, waste from the estimated 30,000 dairy cows housed in the many factory farms along the interstate has contaminated the groundwater. Locals also speak of the fine mist of manure that often hangs in the air, a factory-farming byproduct that has been found to cause asthma.

New Mexico, moreover, is one of the most "food-insecure" states in the nation. As recently as 2007, it was ranked as the number one place in the nation where people lacked reliable access to adequate food. One in every seven persons struggles with hunger in the state.[3]

On the other side of the food issue, the state also struggles with rising obesity and diet-related disease. In 2011, the state's adult obesity rate was 25.6 percent, up from 11.6 just fifteen years before—making it one of seven U.S. states to experience a doubling of obesity in that time.[4] Its diabetes rate as of 2011 was 8.3 percent—up from 5.3 percent.[5] Not coincidentally, many local residents who once made their own nutritious meals no longer cook, let alone grow, their own food. And, as if that were not enough for the young leaders trying to change all this, the three leaders of La Semilla are not even farmers.

"Do you think we're crazy?" asks Dominguez-Eshelman. "We ask ourselves if we're crazy sometimes."

Crazy or not, they are attracting significant support for their efforts. In 2011, the W. K. Kellogg Foundation awarded La Semilla a three-year grant of $432,680. Kent Halla, owner of Sierra Vista Growers, the largest nursery in the Southwest, gave them fourteen acres of land to farm. And Olga Pedroza, a prominent city councilwoman, is just one of the community leaders who routinely champion their cause. The reasons are threefold: the critical needs they have identified, the solutions they are proposing—and the emotional, social, and ecological intelligence they are bringing to the effort.

FROM FARMS TO FACTORY FARMS

The system La Semilla is tackling has changed dramatically since the 1960s. As Roni Neff of the Johns Hopkins Center for a Livable Future summarized,

> In the last fifty years, vast tracts of land devoted to corn and soy have largely replaced farms that raise animals and grow fruits, vegetables, and a variety of grains. Animals raised for food or to produce food have been moved from farms to feedlots and confinement operations. And instead of family farmers, large corporations such as Tyson, Smithfield, Cargill, and ConAgra control much of the process and market.[6]

These changes stemmed from what originally seemed to be a very good idea: the application of scientific breakthroughs to increase agricultural yields dramatically. Indeed, the practices of the "Green Revolution" earned Norman Borlaug, the plant scientist whose work inspired it, the Nobel Peace Prize in 1970.[7] And corporations and governments, recognizing the opportunity presented by the new technologies, fostered the rapid spread of industrialized agriculture from the United States to Asia and Latin America.

On the face of it, industrialized agriculture promised to be a most welcome solution to the timeless problem of world hunger. But some so-called solutions, as writer and farmer Wendell Berry observed, led to ramifying sets of new problems.[8] And during the past several decades, it has become increasingly clear that industrial agriculture has indeed created a host of new problems impacting the health of people and the planet. The use of fertilizers and pesticides, for example, has led to higher rates of cancer and the contamination of soil, streams, and groundwater. Monoculture farming (the cultivation of a single crop over a large area) has led to the loss of biodiversity, undermining the productivity and stability of ecosystems. Factory farms, where most chickens, hogs, and cattle are now bred and slaughtered in the United States, contaminate water and soil and create air pollution linked to asthma and other respiratory problems.

Modern agriculture, moreover, is the single largest user of water worldwide, with global agriculture consuming nearly two quadrillion gallons of rainwater and irrigation water annually—enough to cover the entire United States with two feet of water, as essayist Wenonah Hauter has reported.[9] Industrialized agriculture is also highly dependent on diminishing supplies of fossil fuels, accounting for some 19 percent of fossil fuel consumption in the United States. It is also one of the most significant contributors to the greenhouse gas emissions that cause climate change.[10]

Meanwhile, the reality of hunger and malnutrition remains. In 2009, the United Nations reported that the number of hungry people in the world topped one billion—one of every seven men, women, and children. In recent decades, there has also been an epidemic of obesity and diet-related illnesses in developed countries such as the United States. The causes are complex, to be sure, but one of the primary reasons is the rise of fast foods and processed foods that are high in fat and sugar and low in nutrition. Because these foods are cheap—often cheaper than more nutritious options—many low-income people have come to rely on them to the detriment of the health of their families and communities in places like southern New Mexico.

HOW TO FEED NINE BILLION

With world population projected to increase from seven to nine billion people by 2050, how to feed everyone—without exacerbating ecological damage—is a big question, especially given that one billion among us are already hungry.

But Olivier De Schutter, the United Nations Special Rapporteur on the Right to Food, issued an encouraging report in 2010 that concluded it is possible to improve agricultural productivity significantly in developing countries, reduce poverty, and preserve ecosystems.

The answer, he said, is agroecology, which applies the science of ecology to the study, design, and management of sustainable agricultural ecosystems. Among the core principles are these:

- Recycle nutrients and energy on the farm.
- Integrate crops and livestock.
- Diversify species and genetic resources over time and space.
- Focus on interactions and productivity across the agricultural system, rather than focusing on individual species.

"Today's scientific evidence demonstrates that agroecological methods outperform the use of chemical fertilizers in boosting food production where the hungry live—especially in unfavorable environments," he said.[11]

Entitled "Agro-ecology and the Right to Food," the U.N. report also found that agroecology improves resilience to climate change with techniques that promote biodiversity, one of nature's basic support systems. By reducing the use of fossil fuels, it also "puts agriculture on the path of sustainability."

FROM NOT KNOWING TO KNOWING

A thirty-minute drive north from El Paso, Texas, and south from Las Cruces, New Mexico, the town of Anthony is divided across two state lines. On the New Mexico side, Anthony has a population of nearly 8,000—with some 97 percent of those who live in "colonia communities" earning less than $5,000 a year, according to the U.S. Department of Housing and Urban Development.[12] Many work at seasonal jobs such as harvesting pecans that are exported to China, or they take odd jobs where they can get them. On the Texas side, Anthony has fewer than 4,000 people and is home to a federal correctional institution; several food stores, such as Big 8 and Circle K Drive-In; and all the usual fast-food restaurants: McDonald's, Burger King, Wendy's, Carl's Jr., and Little Caesars Pizza.

Xavier Hernandez, a high school student from Anthony, remembers when no fast-food restaurants existed in his community. But as they became ubiquitous, many people became accustomed to eating fast food without a second thought about nutrition. "I didn't know much about food or nothing," says Christopher Garcia, a young man meeting with friends in Anthony's community garden. "I'd eat fast food or whatever. You know, I'd just go out and eat." His sister, Priscilla, who holds a baby in her arms, agrees: "I didn't know about growing. I didn't even know there was organic food."

But when they heard about a program that would pay them to work in a garden and learn about healthy food, they signed up—just for the chance to make some money. What they didn't expect was that the experience would turn their lives around.

"I learned how to have a relationship with my food and be more mindful about everything I do," says Priscilla Garcia. "I learned that where we get our food is so important, and how everything is so connected, and just how food itself nurtures your body and how we nurture it." With tears in her eyes, she adds, "Just the whole manifestation of my life, I believe, happened here."

While working at Colonias Development Council, which funded the community gardens project with support from a National Park Service grant, Dominguez-Eshelman, Sharratt, and Wiggins-Reinhard offered classes; engaged young people in the Youth Food Policy Council; took them to events; and connected them with growers and food activists—efforts they continue now through La Semilla. In the process, they work hard to get to know the participants individually, believing that relationships are key to any authentic learning experience.

They also take care to be nonjudgmental. For example, they tell participants in their programs, "Do not go home and tell your parents that we just said that everything you buy is wrong, and you should be buying organic," explains Dominguez-Eshelman. "But we do say you have a right to know that there are differences in food. And if you want that access [to healthy food], you have a right to ask why you don't have it, and how can we work towards it."

Manuel Garcia, brother of Priscilla and Christopher, is planning to also join in the work of La Semilla. As he has become more aware of the importance of eating healthy food, he says he has become increasingly, passionately committed to educating others. "I don't think it's a coincidence that minorities, in particular poor minorities, have higher rates of diabetes and heart disease," he says. "And I really want to start educating people on the negative health effects that a lot of the food they sell us has on our lives—not only on our lives, but also on our families, on our communities."

One discovery that particularly influenced him followed some research into the chemicals that go into fast foods. He found, for example, that American McDonald's French fries and chicken nuggets are cooked in oil that contains dimethylpolysiloxane, a form of silicone that prevents hot oil from foaming.[13] "It is the main component of Silly Putty. It's basically like a rubber component. It's also used in caulk that plumbers use to make the pipes watertight and in all sorts of medical equipment. So I started to think, why is this in my food?"

Test Your "Food IQ"

1. **Which food choice is healthier?**

   A. Chocolate Cheerios

   B. Quaker Natural Oats, Honey & Raisins Granola Cereal

2. **Which food product has fewer ecological impacts?**

   A. Nesquik Strawberry Milk

   B. Horizon Organic Milk

3. **Which food company is more socially responsible (through, for example, its treatment of employees)?**

   A. Green Giant

   B. Eden Organic

If you guessed Quaker, Horizon, and Eden, you might count yourself "greenwashed." Greenwashing creates the illusion that a product is entirely healthy or ecologically friendly by featuring a few positive attributes—and ignoring many others.

But you can find accurate information about more than 150,000 food, personal care, and other products by consulting GoodGuide.com, a website and mobile app that provides information about the health, environmental, and social performance of products and companies, using a simple 0 to 10 rating system.

Founded by Dara O'Rourke, an industrial ecologist and professor at the University of California, Berkeley, GoodGuide relies on a team of chemists, toxicologists, nutritionists, sociologists, and environmental life-cycle assessment experts to gather and evaluate data from more than 1,000 different sources. Food products, for example, are evaluated on the basis of their levels of sugar, sodium, cholesterol, saturated fat, and trans fat; and the presence of potentially hazardous food additives, genetically modified organisms, and high-fructose corn syrup.

Ultimately, explains O'Rourke, the goal is to use this information to inspire improvements in food systems. "The idea is that by providing better information," he says, "you empower people to vote with their dollars, express their preferences, and exert influence over brands and retailers."

These are the kinds of stories that convince the leaders of La Semilla of the potential for transforming individual and community food practices through education. "Just learning about that bigger picture and seeing how connected they are to it—and that they still have power, by the choices that they make or don't make in terms of what food they buy. All of us have the power to make changes even at that small level," says Dominguez-Eshelman.

There is also a deep sense of belonging—to the families, to communities, to nature, to life—that comes from reengaging in food at the community level, says Sharratt. "Unless you're involved in the food system or involved in growing, everybody becomes so disconnected from where our food comes from, what it means for our bodies. Now you go into a grocery store and there's no connection whatsoever. It's all covered in plastic." La Semilla is trying to help people rediscover those connections by showing them where food actually comes from, he says, and helping them recognize the impact of food—for better or worse—on people's physical and emotional lives.

CHANGING THE CLIMATE

Ricardo Salvador, formerly a professor of agronomy at Iowa State University and now a program officer with the W. K. Kellogg Foundation, grew up near Anthony. He knew it to be a conservative place, deeply invested in the commodity production of cotton and pecans and holding onto its factory farms. He also knew that its residents suffered from food-related health problems. That's why one of the first things he did when he joined the Kellogg Foundation in 2006 was to search for an organization interested in inspiring a change in the region's food system. But although he searched widely, the promise of grant money in hand, he could not find any takers. "I ran into everything from confusion, to bewilderment, to opposition—everything but enthusiasm," he recalls.

During the past few years, however, interest in food and health has begun to grow in the region, in part because of the New Mexico Collaboration to End Hunger, a statewide public-private partnership formed after New Mexico was declared the most food-insecure state in the nation, and in part because of the local efforts of Dominguez-Eshelman, Sharratt, and Wiggins-Reinhard.

"Two years ago, people weren't anywhere near as ready as they are now for change" in the food system, says Olga Pedroza, city councilor for Las Cruces, New Mexico. "And I think that now, because of all the things that [the La Semilla founders] have done—the food summit, and the teaching, and the presentations, and the relationship building—the climate of the place is beginning to get a little bit more sympathetic."

Salvador is even more enthusiastic. "I never thought the work they are doing would take root in that place," he says. "But they have brought together a really broad coalition of people with convergent interests," including government officials, businesses, farmers, and educators. "I really respect that they have already gathered significant resources—and I am not referring just to financial resources, but significant social capital in support of the alternative vision they have."

Salvador admits that the Kellogg Foundation views La Semilla as a high-risk investment: The leaders are young, the organization is a start-up, and their vision is daunting. But Salvador believes their plan makes sense, and they have the people skills to build a broad base of community support.

Dominguez-Eshelman, for example, "clearly thinks both deeply and far in the future and has the emotional intelligence that allows her to lead from behind," Salvador says. "She works with the process of the group. She doesn't get ahead of collective thinking. She gently guides in a particular direction. And what makes her particularly effective is that she doesn't have a predetermined agenda with others. I think she literally is one of the best listeners I've ever run into. She listens, processes, and then puts her thoughts out there."

The three founders are also effective collaborators, because each brings an important set of skills that the others value, and no one person tries to overshadow the rest. Sharratt, for example, is appreciated for his ability to express the group's ideas in writing—most notably, in grant applications. And Wiggins-Reinhard is valued for being able to jump into action and make things happen with great enthusiasm for the young people with whom they work.

"I never thought I would enjoy working with teenagers, but I found that I love it," Wiggins-Reinhard says. "We interact with youth and their families on such a personal level; it is just incredible to watch them grow and transform into young adults and advocates for change in their communities."

FROM EDUCATION TO ACTION

But if real change is to come to the food system in southern New Mexico, it will take more than good ideas, people skills, and even coalition building. That is what led the founders of La Semilla to the idea of starting a youth farm—a place where young people can learn about sustainable farming, permaculture, nutrition, culinary skills, and entrepreneurship.

"I think the farm is so integral," says Sharratt, the son of a soil scientist with the United States Department of Agriculture. "That is the piece of it for us that we're so passionate about focusing on. I feel like part of that comes out of this recognition that for any change to happen—this desire for farmers' markets everywhere—we need to have production. Because we can have these conversations until we're blue in the face. But if people aren't producing food for local markets, well..." he pauses there, as if to say, "...then it is just talk."

They know, of course, that the food they will grow on the land they only recently acquired will not be sufficient to feed the whole region. But they want the modest farm to serve as a demonstration site to show others what is possible—that young people can grow their own food, even in the desert. And you never know what is possible after that.

They never imagined that they would be able to acquire land in the first year of their organization's existence, for example. But Halla, who appreciates their goal of showing people a positive alternative, gave them fourteen acres outright when he learned what they were doing. Walking across his own property (he owns some 180 acres), the sixty-five-year-old nurseryman says, "I just have a good feeling about them, and if they're interested in doing this, I'm interested in helping them."

To learn more about what it takes to run a farm, Dominguez-Eshelman recently attended the University of California, Santa Cruz, to earn a certificate in ecological horticulture from the Center for Agroecology and Sustainable Food Systems. She was glad for the chance to accelerate her learning curve, but she recognized at the same time that the more she knew, the more she didn't know. So, true to their collaborative spirit, La Semilla's founders will not be diving into farming alone. As of early 2012, they formed an advisory committee to help them develop a farm plan and hire a farm manager.

Behind them are a number of powerful players committed to ensuring that they have every possible chance of success. "We have a lot staked on this," says Salvador. "We're not leaving it up to chance. We're doing everything we can to support them." The Kellogg Foundation is, for example, providing technical assistance and helping the organization network with groups in higher income communities in northern New Mexico, such as Farm to Table, a Santa Fe–based nonprofit dedicated to promoting local agriculture.

If La Semilla lives up to what many believe to be its potential, Halla could eventually turn his entire business over to the organization in the form of a foundation, a prospect he has begun discussing with Salvador, Kellogg's program officer.

But for now, the leaders of La Semilla are remaining grounded in the English translation of their name: the seed. They understand their mission to be seeding actual food—and the idea of nutritious food for all. "In a way, it is about planting those seeds so that we're not the only ones speaking about the region and what the needs are," says Dominguez-Eshelman.

In the end, how else are people going to change a complex system controlled by corporate and government interests? Halla adds, "You've just got to start one person at a time, one area at a time—and seed."

Looking back on this story, think about the ways in which the network of relationships among community members was strengthened through the La Semilla Food Center.

Forging the Food Justice Path

Chapter
EIGHT

Tony Smith, Superintendent of Schools, Oakland, California

In this story, oakland school Superintendent Tony Smith shares his vision of a "full-service community school district" that provides an array of services to students and their families so that all children have an equal chance to thrive. As you read, consider how his own life experiences influenced his emotional and social intelligence, and how he integrates these skills with a capacity for systems thinking, a core dimension of ecological intelligence.

IMAGINE STANDING OUTSIDE a hospital nursery window, looking at two newborn baby boys side-by-side in their plastic bassinets, fists curled and screaming their little lungs out. Both appear healthy, full of life, and, you imagine, ripe with potential. Yet the reality is that after the boys arrive at their homes, the disparity in that potential will soon become evident. Because of the difference in where they live—which is just one ZIP code number apart—one of those babies will grow up to eat healthy food purchased from a supermarket or farmers' market. He will finish high school, attend college, and live to the age of eighty-two. The other boy will fill up on unhealthy food purchased from a corner liquor or convenience store. He will drop out of high school. And he will die fifteen years before the first.[1]

These are the everyday facts of life in Oakland, California, the third most diverse city in America, but they are not unlike those in many other urban areas across the nation. Food is just one factor, but in recent years, there has been a growing awareness that inequities in food access have a significant impact on children's education, health, and life expectancy—and that a radically new approach toward food is needed to ensure equity for all schoolchildren.

One of the people leading the effort to create such an approach is Tony Smith, a former San Francisco 49ers football player who is now superintendent of the Oakland Unified School District.[2] Since he took over the district in 2009, Smith has been working on a comprehensive plan to transform Oakland into one of the first "full-service community school districts" in the nation. In Smith's vision, a full-service community school district weaves together a broad network of services—through government agencies, the business community, foundations, philanthropists, and nonprofits—to level the playing field for all children.

He is, for example, working with the Alameda County Health Department to create school-based health care centers that offer services that range from administering vaccines to counseling children who have lost friends or family members to violence. He is encouraging the local business community to create internships for young people who come from families in which no one has held a job for generations. And he is striving to improve the quality of food Oakland students eat—both in the school cafeteria and at home.

As someone who struggled against the odds himself, Smith is not naïve about the seriousness of the challenges he faces. Oakland, true to its reputation, is a tough place to live. Gang violence, racial tensions, and conflicts between police and residents date back to well before the founding of the Black Panthers there in the 1960s. Oakland suffers from high rates of poverty and unemployment, as well as a history of inept city government. The school district itself, which was placed under state supervision due to high debt and a record of mismanagement before

THE RISE OF SCHOOL FOOD REFORM

Once a seemingly fringe issue embraced in places such as Berkeley, California—but ignored or even scoffed at elsewhere—school lunch reform has become widely recognized in recent years as vital to health, education, the environment, and the economy. First Lady Michelle Obama made it central to her plan to solve the problem of childhood obesity within a generation. Numerous schools have taken it upon themselves to make improvements. And celebrity chef Jamie Oliver highlighted it as the focus of a reality TV show in 2010.[3]

One of the biggest reasons for this growing attention is the realization that one out of three American children is overweight or obese, due primarily to unhealthy diet and lack of physical activity. This fact presents significant public health and economic hazards, threats, dangers, and challenges. When children are obese or overweight, they tend to remain so throughout adulthood—putting them at increased risk of heart disease, diabetes, hypertension, stroke, osteoporosis, and cancer.

Treating millions of people with diet-related diseases places an enormous burden on an already fragile economy. The amount of money the United States spent to treat obesity-related conditions doubled between 1998 and 2008, rising to an estimated $147 billion per year in 2008.[4]

the young man known for being a kind-hearted defensive lineman who helped his teammates with their English assignments signed with the Green Bay Packers and later the San Francisco 49ers. Injuries forced him to leave both teams; by the time he was twenty-four, the football career he had planned as his pathway out of poverty was over.

In an effort to regroup, Smith considered studying law. But Jo Baker, the director of UC Berkeley's Athletic Study Center at the time, strongly suggested that he consider teaching. "He was always interested in others' development, not just his own," says Baker, who had known Smith since he was a freshman. "He was able to break things down, make things easy for people to absorb. And he's an optimist. To me, that's a teacher."

Convinced that Baker was right, Smith returned to Berkeley, where he earned a Ph.D. in language and literacy from the School of Education. This experience, he reflects, helped shape his approach to reform in Oakland, because it helped him understand the history of education and the reasons why schools are designed the way they are. Above all, his studies helped him recognize how schools reinforce cultural and class inequities: Even the notion that only a few make it to the top, he says, serves as "a powerful way to keep things where they are."

Smith is, of course, one of the few people with such a background who found success as an adult. But although his determination and well-rounded intelligence certainly contributed to his success, so, too, he insists, was the support he received from educators, coaches, and other individuals along the way. "I wasn't cared for in the ways I should have been, but I was cared for by people who chose to care," he says. "And I know for a fact that if not for other people—a coach or teacher who said, 'Come here, you need to stop doing what you're doing; you have worth; you have skills; come this way'—I would not be here today." That is the kind of transformative support he wants to offer now, not just to a few individual students, but to a whole school district situated in a community that has more than its fair share of challenges.

SURFACING SYSTEMIC INEQUITIES

Oakland is the sixth most dangerous city in America, according to the FBI.[12] In 2011, an average of five to six people a day (more than 2,000 people per year) were victims of gun violence. And, as of August of that year, 199 of those shooting victims were children. That's up 60 percent from 2006, according to the Urban Strategies Council.[13] Some residents blame the proliferation of handguns for the violence. Others fault an understaffed police department. Still others say the problem is the city's high unemployment and poverty rates. In 2011, Oakland's unemployment rate was 16.5 percent,[14] almost twice the national average.

At the same time, as Smith enthusiastically points out, Oakland is also a dynamic city with many beautiful neighborhoods, an extensive park system (11 percent of

city land is dedicated to parks), trendy new restaurants, and a burgeoning arts scene that earned it the fifth spot on the *New York Times* list of "The 45 places to go in 2012."[15]

What explains the two faces of Oakland? Smith and other leaders point to systemic inequities in race, class, and ethnicity that affect such fundamentals as food, health, and education. For example, in West Oakland, which is populated largely by poor African Americans and Hispanics, there is only one supermarket for every 93,126 residents, according to the Hope Collaborative, an Oakland-based organization focusing on environmental health and food policy issues.[16] And many residents can't travel to the supermarket because they don't have access to a vehicle or public transportation. So they buy their food at small stores within walking distance of home—liquor stores or convenience stores that rarely carry fresh, nutritious, affordable fruits and vegetables amid the shelves of unhealthy, highly processed (and often highly priced) food options.[17]

Higher income whites living in North Oakland, in contrast, have one supermarket per 13,778 residents, according to the Hope Collaborative. They also have access to several farmers' markets, where they can buy organic produce in season, while enjoying samples of healthy food and live music.

Stark health differences develop in adulthood, largely as a result of these disparities in food access. An African American resident in West Oakland, for example, is five times more likely to be hospitalized for diabetes than a white resident born in the more affluent Oakland Hills. Moreover, that same African American resident is two times more likely to die of heart disease or cancer as the white resident.[18]

Not surprisingly, significant disparities in academic achievement also exist in Oakland. According to the Oakland School District, only 54 percent of African Americans and 56 percent of Latinos graduate from high school, while 75 percent of white students and 79 percent of Asian American students graduate, with many going on to college.[19]

And the racial achievement gap is clearly not limited to Oakland. "It's a pattern that plays out city by city, district by district, state by state, across the country," Smith told a recent Chamber of Commerce meeting. "The system is producing the specific and exact outcomes that it is designed to produce if it's this consistent across all districts. . . . It's a tremendous waste of human potential to keep sorting and selecting the way we do, and to have this kind of predictable gap over and over and over again."

Although poverty and other factors are surely to blame, the racial achievement gap is also increasingly being recognized as connected to the poor nutritional quality of food available to many African American and Latino students in Oakland and elsewhere. And that is one of the things Smith is trying to change in Oakland—in more ways than one.

If your school eliminates chicken nuggets and offers a salad bar instead, will students change the way they eat outside of school? Not necessarily—if that's all you do. But if you also integrate classroom lessons about healthy eating and hands-on cooking or gardening classes, you will have a much better chance of changing students' knowledge, attitudes, and behavior toward food.

That was the finding of a study of the School Lunch Initiative, a pioneering collaboration among the Center for Ecoliteracy, the Chez Panisse Foundation, and Berkeley Unified School District, which was one of the first comprehensive school lunch reform efforts in the nation.

Conducted by the Dr. Robert C. and Veronica Atkins Center for Weight and Health at the University of California, Berkeley, the study found that in those schools that combined healthier food, class-room instruction, and cooking and gardening classes, the following occurred:

- Sixty percent of families surveyed said that the school changed their child's knowledge about healthy food choices (in contrast to thirty-six percent in schools that only changed the food).

- Forty-two percent said that school changed their child's attitudes about food (compared to nineteen percent in schools that only changed the food).

- Thirty-five percent said that school improved their child's eating habits (versus sixteen percent in schools that changed only the food).

To read the report, "Evaluation of the School Lunch Initiative: Changing Students' Knowledge, Attitudes, and Behavior in Relation to Food," visit the Center for Ecoliteracy website at www.ecoliteracy.org/downloads/school-lunch-initiative-evaluation.

DRUGS THERE, APPLES HERE

On an unseasonably warm afternoon in January, a group of students filed out of a high school in East Oakland—home to an area known to have the highest rates of shootings, prostitution, and drugs in the East Bay. Rosa Arciniega, a Latina mom with a white apron tied around her waist, approached the first group. "Fresh apples," she cheerfully called, directing them to some nearby tables filled with apples, broccoli, chard, and a dozen other items under a white canopy. A few students half-heartedly walked over. Then history teacher Alfredo Matthews appeared, pulling out his wallet, waving a dollar bill above his head, and firing off questions to the newly alert group of students: "What's beneath the atmosphere? Who can tell me how many teeth are in the human mouth? What came first: the chicken or the egg?" Every time a student gave a correct answer, he gave them a dollar bill. "But you have to spend it at the farmers' market," he called out.

The farmers' market at East Oakland's Castlemont Community of Small Schools is held every Wednesday afternoon, inside the front gate. Only a few years old, it sells primarily to teachers at this point, says Kelly Quane, a high school English teacher who volunteers to bring the produce to the school. Many of the kids don't have the money to purchase food, she says, which is why her colleague, Matthews, made a game out of giving away money so a few kids would experience the taste of a fresh apple. Quane hopes that local residents, even just those who live across the street, will try the produce, because the prices are more affordable here than in the supermarket more than a dozen long blocks away. "But it's a hard community. We can't go knocking on the doors," she says, adding that they will try leaving flyers on doorsteps.

Still, the kids see the fresh food there every week. Many receive tastings from Arciniega, whose own children are grown, and who can't seem to resist giving out samples. And some get a good deal more involved than that. Melinda Monterroso, a senior who calls herself a "green pioneer," is one of the East Oakland students who regularly volunteer to work at the stand. "My dad got sick because of the way we were eating. It was unhealthy. That was a wake-up call for me," she says. "Having it here at the school is a reminder to eat healthy," she adds, flashing a big smile.

Senior Omar Mateo says he became interested in healthy eating because of a teacher who engaged him in a school garden project and taught him about food. "She told us it was not healthy to eat chips all day. Just because it tastes good doesn't mean it's good for us." He volunteers at the farmers' market because that teacher (who has since gone back to school herself) taught him to take care of his

RETHINKING SCHOOL LUNCH

The Center for Ecoliteracy's Rethinking School Lunch planning framework is based on a positive vision: healthy children ready to learn, "food literate" graduates, invigorated local communities, sustainable agriculture, and a healthy environment.

The framework identifies ten pathways for innovation and change in school food. Because they are interrelated, school food reformers can begin with any of them, knowing that the change process will eventually lead them to the others. They are:

- *Food and health:* The school food program promotes student health by offering nutritious and appealing menu options.

- *Wellness policy:* The school nutrition program is guided by a district wellness policy developed through a collaborative community process.

- *Teaching and learning:* Hands-on learning, the lunchroom experience, and teaching and learning in the classroom deepen students' knowledge, skills, and attitudes about food, culture, health, and the environment.

- *The dining experience:* Students feel welcomed, safe, and valued in an atmosphere that encourages healthy eating and positive social interaction.

health. He also likes that Arciniega shares leftovers with the student volunteers when it is time to close for the night.

CHANGE: FROM PIECEMEAL, TO SYSTEMIC

Oakland schools had made a number of improvements in school food before Smith came on the scene. This was due largely to the dogged determination of district food service director Jennifer LeBarre, who recalls earlier days when she poured nacho cheese sauce into a bag of Flamin' Hot Cheetos and called it lunch. But since then, the district has banned soda (even before the state mandated it), trans fat, deep fryers, high-sodium foods, and most white bread products. It introduced "meatless Mondays," universal breakfast, daily fresh produce at every school, increased fresh food prepared onsite, and local purchasing.

In collaboration with several community organizations, Oakland Unified also organized weekly farmers' markets like the one at Castlemont at twenty-two of its ninety-one schools—mostly in communities known as "food deserts," where there is little access to fresh produce. With the help of the local food bank, the district offers a program that sends some of the poorest students home on Friday afternoons with backpacks filled with healthy food to last the weekend. It also provides a mobile food pantry at two schools twice a month, which attracts hundreds of students and parents. Salad bars now operate in sixty-two schools; some are so popular that strawberries and other fresh fruit have to be rationed.

- *Procurement:* The school meal program buys fresh, seasonal, sustainably grown food from local and regional sources.

- *Facilities:* The dining facility serves as a learning center, offers fresh food prepared onsite, and reinforces lessons learned in the classroom.

- *Finances:* Wise budget planning makes the shift to fresher, more nutritious food financially viable.

- *Waste management:* The school food program reduces waste and helps students understand the need to conserve natural resources.

- *Professional development:* Nutrition services staff and teachers receive the training they need to prepare healthy food.

- *Marketing and communications:* Districts promote healthy meal programs and meaningful learning environments.

You can download the Rethinking School Lunch guide on the Center for Ecoliteracy's website at www.ecoliteracy.org/downloads/rethinking-school-lunch-guide

Still, there are limits to what even dogged determination can achieve when a school system faces the kind of structural obstacles confronting Oakland and many other districts. Oakland schools, for example, serve some 36,854 meals a day—on a very slim budget (with a maximum reimbursement from the federal government of $2.77 per lunch and $1.51 per breakfast).[20]

Fewer than one in four Oakland schools has a working kitchen. Even in schools with kitchens, most of the equipment is so old that it no longer works. Many of the 300 food service workers do little more than rip the plastic off pre-packaged food and pop it in the microwave. Real change to provide healthy food for all students would require changing a complex and deeply flawed system.

That's why Smith approved a collaboration between Oakland Unified and the Center for Ecoliteracy, a pioneer in school lunch reform for more than a decade, to develop a comprehensive Rethinking School Lunch Oakland feasibility study about reforming the district's school food environment.[21] The aim of the study was to determine what it would take to transform Oakland school food, based on the model of the Center's ten-point Rethinking School Lunch Initiative, which addresses every step of the process from procurement to waste.

In addition to a thorough analysis of the obstacles to improving the quality of food served to the district's 38,000 students—70 percent of whom are eligible for free or reduced-price lunches—the study made a number of recommendations. Among them are the following:

- Develop a 1.5-acre farm and central commissary kitchen in West Oakland.
- Refurbish seventeen cooking and fifty-eight finishing kitchens across the district to move away from pre-packaged meals and toward cooking with fresh ingredients.
- Transform ordinary school kitchens into "school-community" kitchens that local residents can use during off-hours for cooking, culinary training, and business development. Says Smith, "Imagine school not only being a place where you go for a PTA meeting, but also where you go to do some cooking with community members—a place where you can have twenty-five, forty people. That starts to change the character and nature of the place. As a community member, you start to feel it's yours."

Achieving these changes, projected to cost $26 to $27 million over five years, will take a lot of effort, given the series of budget cuts the district has withstood, the need for significant facility upgrades to meet seismic requirements, and public controversy following the superintendent's decision to close some schools in order to become more cost-efficient.

Still, Smith is confident that the community will support healthy, local food in ways that nurture families, communities, and the environment for all students. "Look at all the food leaders who are already here in Oakland," he says. Among them are People's Grocery, a food justice organization based in West Oakland; Revolution Foods, a for-profit firm that offers healthy, freshly prepared meals to a growing number of schools nationwide; Kaiser Permanente, one of the largest not-for-profit health plans in the nation; and the Oakland office of The California Endowment, a public foundation that announced in 2010 that it will dedicate $1 billion during the coming decade to improve the health and health care of Californians.

Children of color are too often blamed for poor outcomes in schools and in life, Smith says, as if those outcomes are determined by their own efforts alone, and not affected by the adults in their lives and the systemic forces that do not provide support or that reduce chances of success.

But truly changing outcomes for young people, especially when it comes to equity issues that have plagued a community for generations, can occur only when those systemic inequities are addressed, and adults throughout a community show genuine empathy for its youth. In this respect, Smith says he is inspired both by his own personal experience and by the work of people such as Father Greg Boyle, who helps former gang members gain jobs, training, and education in Los Angeles.[22] Drawing from the words of Boyle (who in turn quoted Mother Teresa), Smith says, "If we behaved as if we belonged to each other, what would we do? We would turn toward each other."

Then he adds softly, "If you really believe that, it changes how you are in the world. And I believe that."

Like many large U.S. cities, Oakland is home to huge disparities in income, neighborhood violence, and access to food, public transportation, and health care services. Students similarly suffer from significant disparities in academic performance. Think about what you have read here and reflect on Tony Smith's belief that "When you move a system, you have to move all of it."

PROFESSIONAL DEVELOPMENT STRATEGIES

Cultivating Ecoliterate
Learning Communities

Guidelines for Engaging Colleagues

T HE PEOPLE PORTRAYED IN this book are community activists, educators, artists, students, and scholars who apply emotionally and socially engaged ecoliteracy to urgent environmental challenges. They embody hope, empathy, and resiliency that can serve as inspiration to us all.

This chapter offers educators ideas for examining the five practices of engaged ecoliteracy described in the book's introduction. Beginning with cultivating ecoliteracy within your community, we share some guidelines that we use at the Center for Ecoliteracy to connect colleagues who are interested in transforming teaching and learning to meet the challenges of the twenty-first century.

The current generation of students has inherited daunting environmental issues and challenges that will require major shifts in conceptual understanding, ways of thinking, and everyday behaviors. Although we are well aware of the overwhelming pressures on educators to cover a myriad of topics with limited time and resources, the Center for Ecoliteracy recognizes that the urgency of our global crises compels the education community to rethink curriculum priorities and to infuse a sustainability perspective into the curriculum.

Of course, that is easier said than done. This chapter offers strategies for getting started at your school and reflects the Center's many years of working with educators to cultivate their emotional, social, and ecological intelligence and, in turn, those of their students, in ways that enliven and enrich education.

FORM A LEARNING CIRCLE

The first step in cultivating ecoliterate learning communities is to seek out like-minded colleagues with whom to work. Although it is possible to apply the practices in this book for individual professional growth, it is well documented that teaching and learning in a school community is enhanced when educators engage with each other. At the Center for Ecoliteracy, we have found that participation with colleagues over time increases the power of sharing wisdom, building trust, and offering support. These engagements can take many forms—two or more full days of intensive professional development, one-hour meetings every other week for a semester, or some other configuration. If possible, it's helpful to involve all potential members of your group in developing the meeting schedule.

We suggest that you experiment with "learning circles." Learning circles are based on a form of meeting that originated among indigenous people. Because the shape of a circle implies equality and inclusiveness, we suggest that you literally sit in a circle. Learning circles are different from a typical faculty meeting or discussion in that all members share an expectation of thoughtful speaking and attentive listening. Also, the pace of a circle is usually slower, allowing time for members to make sense of what they have heard before hearing or speaking additional thoughts.

We'll use the term "learning circle" throughout this discussion, but our suggestions can be adapted for the format you prefer, naming it by whatever term makes sense.

A learning circle can be as small as two participants or as large as your entire faculty and preferably comprises people who recognize its value. Ideally, participation is voluntary, and all members are enthusiastic about their membership. However, professional development experiences rarely attract the same level of commitment from everyone (and are sometimes mandatory), so inviting input from members for how the learning circle will work will increase its likelihood of success.

Through the Center for Ecoliteracy's work and feedback from educators, we have identified several elements that we incorporate into our professional development gatherings. These elements help create an atmosphere that is intellectually based, yet heartfelt. They may provide food for thought for you and your colleagues to design a learning circle that will work for your group. The first set of elements includes four strategies for cultivating ecoliterate learning communities that can serve as a focal point for your learning circle. The second set of elements identifies some logistical considerations that you and other members will want to consider before the circle begins.

USE FOUR STRATEGIES FOR CULTIVATING ECOLITERATE LEARNING COMMUNITIES

The four techniques that follow will get you started by increasing your collective knowledge and practices of ecoliteracy and offering ways to apply them both personally and professionally. Although many effective professional development methods can be used along with this book, we highlight these four strategies because they are especially effective for forming a network of relationships, which is essential to becoming ecoliterate. These techniques are also useful when bringing together students, the school community, and the wider community to discuss topics relevant to creating a sustainable and resilient society.

Personal Reflections

Personal reflections help learning circle members clarify their individual values, attitudes, and opinions before "going public" with their thoughts. They are especially useful prior to sharing views on a sensitive or controversial topic. By taking time to think before speaking, strategizing constructive ways to address conflicting ideas, and preparing for emotional self-management, learning circle members practice key competencies of emotional, social, and ecological intelligence.

Consider a topic from this book, discussed in the sidebar "The Rise of School Food Reform" in Chapter Eight. Some people believe that schools need to maintain a single focus on academics and that spending time and money improving school food is inappropriate, especially in the current economic climate. This may be a topic that circle members can reflect on individually, before expressing their thoughts and opinions in the group.

Personal reflection can be encouraged through a variety of approaches. For example, asking participants to jot down their initial ideas and feelings prior to opening up a topic for discussion increases the quality of discussion and also frees everyone to listen deeply instead of mentally crafting their individual responses while others are speaking. Having drawing paper and colored pencils available allows for self-expression in a visual medium, which can be useful for shifting participants' perspective or seeing through a new lens. Creating collages, working with clay, creating maps, and using visualizations are also imaginative ways to explore new ideas from a personal perspective.

Structured Conversations

Structured conversations offer provocative questions or themes to explore with colleagues. They help build a learning community while simultaneously expanding the learning circle's understanding of ecoliteracy. Frequently, structured conversation follows personal reflection.

A structured conversation usually starts with an open-ended prompt that sparks thoughtful discussions and uncovers ways to apply new ideas to your school setting. For example, after reading Chapter Two, "Taking a Power Trip," your circle might choose to discuss this prompt: The Spartanburg teacher took his students to the power plant where coal was burned, helping to make the invisible visible. How could you help reveal what seems invisible to your students, focusing on an issue of local concern?"

When planning a structured conversation, choose techniques to maximize participation and depth of the discussion and generate diverse and creative ideas. This could mean starting out in pairs before opening discussion to the whole circle, passing a talking stick to ensure that each person has an opportunity to speak, or having members move around the room to share thoughts with several people, one-on-one.

Collaborative Lesson Design

Collaborating on lesson design engages educators in practicing their craft together. Lesson design is at the center of the creative process of teaching, and the chance to participate with others can stimulate creativity. Most teachers, however, have few opportunities to explore ways to teach specific concepts or skills collectively.

Collaborative lesson design can be as expansive as imagining how to convey a particular idea across several grade levels. For example, many of the stories in this book illustrate the capacity to persevere as an important element of engaged ecoliteracy. Learning circle members could generate ideas for learning experiences that foster perseverance at various developmental levels and identify some building blocks throughout a span of grade levels that would continuously develop a student's ability to practice perseverance.

Collaboration can also take on a narrower focus, such as learning circle members deciding to plan one particular cooperative teaching project. For example, after reading Chapter Seven, "Changing a Food System, One Seed at a Time," teachers might decide to take their classes on a joint field trip to a local farm. Learning circle members would use their time together to design lessons for before, during, and after the field trip.

Teaching Rounds

Teaching rounds emulate the professional practice of doctors and therapists, who jointly talk through difficult or intriguing cases or clients. When applied to schools, educators come together in teaching rounds to assess the efficacy of a lesson, a project, or a classroom practice and discuss ways to improve it. Or they can discuss individual students in the context of developing their ecoliteracy.

Most teachers have little time or support to reflect on the success of a particular teaching practice, and the school culture rarely supports a teacher taking time to analyze and improve teaching practices. Instead, teachers usually rely on a gut reaction that indicates whether something they tried went well or didn't go so well. Teaching rounds provide a structured format for thoughtful reflection on ways to improve teaching and learning.

Teaching rounds can also provide time and space for teachers to draw upon their collective insight and experience to address particular issues. For example, some environmental issues—such as climate change—are controversial, and students and their families might hold beliefs and opinions that are at odds with the latest scientific evidence. The introduction to this book, as well as Chapter Six "From Restoration to Resilience," address some climate change issues that may be of interest to learning circle members.

Teaching rounds offer educators a forum for strategizing how best to address these kinds of conflicts while maintaining respect for diverse ideas and, at the same time, staying committed to evidence-based information.

DESIGN A LEARNING CIRCLE FOR SUCCESS

Through the Center for Ecoliteracy's many years of working with and seeking feedback from educators, we have identified several elements we incorporate into our professional development gatherings. These elements create an atmosphere that is intellectual, heartfelt, and practical; you may find them useful for your learning circle.

Agree on Leadership

Learning circles can employ a variety of models for planning and facilitating their gatherings. Although learning circles are by definition run by shared leadership, most educators will not have time to plan every circle together. One or two individuals, who may or may not be circle members, can plan and facilitate the circles; a small subset of the membership can share responsibility for planning and facilitating the whole group; or planning and facilitation can rotate among group members. These approaches prove most effective when all circle members have opportunities to provide input to the group's design, format, and selection of leaders.

Establish Norms

The first learning circle meeting is an optimal time to agree on some ways of relating that maximize participation and trust. Members of the circle are the best people to generate their group norms. Here are some examples of common group norms:

- Show up on time and stay until we adjourn.
- Practice respectful and attentive listening.
- Stay on topic, noting what has relevance to the conversation.
- Participate fully.
- Monitor your talking time.

It can be difficult to remember a long list of norms. After the group has generated a list, take time to identify five to seven most important items that you all agree are necessary to function as a healthy group.

Lucy, Victoria, George, and Vernard—a group of New Orleans students—interviewed Jane Wholey about how she uses learning circles to inspire all students to share their voices. Wholey is founder and executive director of the Rethinkers: Kids Rethink New Orleans Schools. Here is an excerpt from their conversation[1]:

Victoria: What does "circle" mean to you?

Jane: Well, it means a different way of holding a meeting. In school, teachers usually sit behind a desk facing students who sit behind their desks, too. When you take away the desks and put everyone in a circle, they suddenly become equals—even the teacher.

Lucy: OK, cool. So how does a circle work?

Jane: I ask the group a question. Instead of a few kids raising their hands and giving answers, we go around the circle and everyone speaks. Sometimes the person talking uses a 'talking piece,' and passes it to the next person when they're done. There are a few important circle rules. The first is that when a person speaks, everyone listens. The second is that if you don't want to speak, you can pass. The third is that only one person speaks at a time!

George: What goes on in the circle?

Create Clear Agendas

If you ask a teacher to name the biggest barrier in their profession, they will most likely say "time." Too often, professional development meetings feel like an enormous waste of precious time. Designing an experience that is worthy of attendance requires tuning into the time pressures of the learning circle members.

Most teachers appreciate a clear agenda that is followed in the meeting. At the same time, many teachers also appreciate built-in flexibility for addressing issues that emerge from the group, even if they are not on the agenda. Sometimes it is difficult to know when to stick with the agenda and when to address an unexpected but important issue. This decision is often best made by the group, although they may prefer to leave those decisions to the circle leader.

Make Use of Rituals

Establishing group rituals can create a sense of community within the group. Rituals often emerge within long-standing groups. For example, the Center for Ecoliteracy worked with an elementary school faculty that occasionally and spontaneously began singing together. This evolved into a ritual marking the beginning and ending of faculty meetings and other gatherings.

Jane: Conversations…about pretty much anything. Everyone's voice is respected. No one laughs at anyone, and each person has an equal chance to say what they think and feel. You might say that the circle is a sacred space. One way we make the circle feel like a sacred space is to add a 'celebration table' in the middle. We put all sorts of cool things on this table, like pictures of loved ones, candles, and objects that are important to us.

Vernard: Why is the circle important?

Jane: In most meetings I go to, a few people speak a lot and most people speak very little. There's a power thing going on when someone takes up all the space in the room. At Rethink, as you know, we believe in 'power among' not 'power over.' Without embarrassing anyone, the circle makes some people step up and others step back. So everyone gets to practice a new skill—either controlling yourself or overcoming your shyness. When shy people speak, you hear some amazing things.

Lucy: So what do you think the circle teaches?

Jane: More than anything else, the circle teaches equality and the art of deep listening. When you learn those things, you can make the world a wiser, gentler place.

Beginning rituals serve simply as a signal that the group members are present and willing to focus on the issues before them. At the Center for Ecoliteracy, for example, each seminar begins by ringing a simple gong or pleasant-sounding bell, followed by a few seconds of silence; then the person who has come from the farthest distance away lights a candle. This is sufficient for invoking a collective presence. Similarly, a closing ritual honors the focus that has been given by the group and signals that their collective work for that session is complete.

Design the Physical Space

It is widely accepted that people learn best when they have an overall sense of comfort and security. Similar to the Reggio Emilia[2] approach to education, the Center for Ecoliteracy believes in every person's right to well-being, and we place that right in the foreground of planning professional development experiences. Starting with the physical space, we try to secure meeting space with natural lighting and fresh air, which lets the outside in as much as possible.

Within your space, you can optimize comfort by using flexible furniture arrangements that will help nurture relationships as well as make room for personal reflection. Create physical "white space" by eliminating clutter and other unattractive elements and keeping ambient noise to a minimum. Applying the theory that "less is more," try to limit sensory inputs to those that are necessary, beautiful, and inspiring.

"Really good teachers know it is not the one-hour lecture that will resonate," says Aaron Wolf. "It's creating a setting where teachers can draw the energy out in the room and students can come to insights for themselves."

In his own effort to do this with geography students at Oregon State University, Wolf applies the "Four Worlds frame"—reflecting on the physical, emotional, knowing, and spiritual needs of his students—one lens at a time.

"The first thing I do," he says, "is look at the physical setting of the classroom and think about, Are the chairs organized in a way that facilitates discussion, or is it set up for a lecture? Is there natural light? Is there fresh air? Then I think about the physical comfort of the students themselves. Do they know it's okay to bring food and drink into the classroom, and to get up and go to the restroom if they have to?

"Then I think about where everyone is at emotionally," he adds. "Have I given them a moment of transition? Am I encouraging them to be present? Is there discord because of other bigger issues?" In the aftermath of the September 11, 2001, attacks on the World Trade Center, for

Provide Nourishment

Encouraging a sense of well-being can also involve offering access to nourishing food. Not only does food add a welcoming sense of shared community, it is also essential for sustaining mental energy. The kinds of foods the Center for Ecoliteracy provides reinforce our teachings about sustainable living. Food that is simple and seasonal embodies a way of provisioning ourselves that nurtures health and minimizes our impact on other living things, the soil, and other elements of the environment.

Seek Wisdom from Outside

In our seminars, we have found that most teachers desire engagement with provocative and motivating ideas that acknowledge their intellect and feed their minds. After all, most education professionals went into teaching because of a passion for learning. The voices of phenomenal "thought leaders," such as those featured in this book, can lift teachers out of the everyday routine of schooling and inspire them to explore what is possible collectively. Through exposure to stimulating visionaries who foresee a future where all members of society are ecoliterate, educators are reinvigorated to participate in transforming their schools. In addition, hearing firsthand from local leaders who are involved in environmental and social justice issues can help connect these issues to the school community.

example, Wolf observed that students were distracted and nervous. But because the tragedy didn't already fit in with their preplanned curriculum, students didn't have the opportunity to discuss their anxiety with teachers—even though it was getting in the way of their learning. So he dedicated a few classes to allowing students to share their concerns before resuming the planned curriculum.

Wolf next moves to the third lens, which is "where all the fun—and most of the work—comes in": knowing. He routinely asks himself such questions as, "Are intellectual connections being made? Are students contributing to the making of those connections?"

Finally, with respect to taking a spiritual view on teaching and learning, Wolf says that he does not mean a "religious" perspective, but rather a sense of transformation rooted in a strong feeling of connection with the rest of the universe. And this, he suggests, is the "on fire" experience that can develop after the first three needs have been addressed in the classroom.

In short, he adds, "There's not much difference between education and facilitation [of conflict resolution]. We're all just trying to heal the world."

Seek Wisdom from Inside

Although wisdom from people outside the group can stimulate educators to think in innovative ways, uncovering the wisdom within the group galvanizes their collective resolve. Structuring frequent opportunities for participants to share their insights, accomplishments, and dilemmas generates a kinship that can help sustain their individual efforts beyond the professional development experience. In this book, the four strategies for cultivating ecoliterate learning communities, as well as the professional development activities described in the next chapter, offer a variety of ways for eliciting wisdom from within. Of course, many other methods can be used to uncover inner wisdom, such as meditation, slowing down to observe without judgment, and journal writing. Your circle members may choose to use these techniques as well.

Enjoy Yourself

Although most of this book addresses serious subjects, learning circles can be enjoyable gatherings, with room for humor, playfulness, and connection. When participants expect an atmosphere that appreciates humor, they will look forward to attending and see value in the lighter side of serious issues. Once you start learning about issues such as offshore oil drilling or water shortages, you will see that these current topics can often be addressed in the world of comedy.

Encourage circle members to bring in related cartoons, YouTube segments, or jokes that might resonate with members.

Singing, dancing, or listening to music can add levity, a deeper connection to an issue, and inspiration to take action. As with humor, making room for music can add pleasure to the circle experience. Countless songs deal with nature and the environment. (For examples, see the Songs for Teaching website at http://www. songsforteaching.com/index.html.)

Teaching requires a lot of mental and physical energy, so taking a few moments to recharge is usually a good idea. Such a change of pace could consist of something as simple as a five-minute stretch or meditation, or participating in a team-building game. Group rejuvenation can occur at any time during the circle and can be especially appreciated when it's spontaneous.

Making simple food together can be an effective way for circle members to get to know each other in a relaxing context. Preparing something as simple as a fruit salad or a plate of cheeses and breads doesn't require much time but can add an informal, communal element that strengthens relationships within the circle.

Revisit the Five Practices

All of these ideas for coming together with colleagues to cultivate your ecoliterate learning community reflect the five practices of emotionally and socially engaged ecoliteracy. These ideas offer you opportunities to develop empathy for each other and for other forms of life; they encourage you to view sustainable living as a community practice; they help you better understand how nature sustains life; they offer ways to uncover what blocks us from seeing the impacts of our behavior by making the invisible visible; and they provide practice with anticipating the unintended consequences of our actions. Although you will choose ideas and processes that will work for your learning circle, keep these five practices in mind as guides for your professional development journey.

Reflection and Practice

Conversation Strategies for Getting Started

I N THIS CHAPTER, THE CENTER for Ecoliteracy offers ideas to stimulate conversations with your colleagues. These suggestions use the four strategies introduced in the last chapter: personal reflection, structured conversation, collaborative lesson design, and teaching rounds. For each story in the book, you might choose one of the four strategies to discuss ideas with your circle. You might decide to use more than one strategy, or you might use these ideas as starting points for generating your own discussion topics.

PART ONE: STANDING STRONG ON A COAL MOUNTAIN

Personal Reflection

Read the Chapter One sidebar, "Why Can't We Go Out and Play, Daddy?" Using a large piece of paper and colored pencils, express your feelings after reading this story. (Try putting yourself in the role of Little Bopper's parent.) On the other side of your paper, make a list of environmental threats that your school community faces that might affect its members' health and well-being. After responding to this writing prompt, gather with colleagues, allow time for volunteers to share their emotional responses, and then discuss ways that you and your colleagues could participate in addressing local challenges with students.

Structured Conversation

Chapter One, "Lessons from a Coal Miner's Daughter," mentions how Wendell Berry points to a lack of understanding about the difference between the long-term value of a well-maintained forest ecosystem and the short-term gain of coal mining. He says that our education system plays a role in the perpetuation of ecological destruction, because it is based on the faulty premise of an economy that externalizes health, environment, and other costs. What does it mean to externalize health and environmental costs?

Berry calls for a shift from the economy to the ecosphere as the basis of curriculum, teaching, and learning. What do you think about Berry's statements? What are some arguments for and against his position? (Suggestion: Ask half of your learning circle to argue in favor of his position and the other half to argue against it.) Brainstorm ways that the curriculum might change if your school decided to implement his ideas.

Lesson Design

With your colleagues, read the sidebar "What's My Connection?" in Chapter Two. Then view the application, "What's Your Connection to Mountaintop Removal?" at www.ILoveMountains.org. Enter your ZIP code to discover how your community is connected to mountaintop mining. Look at the map that links your community to its source of electricity, and read through the details that follow.

Then, starting in small groups by grade level, generate a list of ways you might use the application with students. (It is most appropriate for students in fourth grade and above.) What do you think they will learn from the application? How do you think they will respond if they discover that some of their energy comes from coal obtained through mountaintop removal?

Obviously, not all students will respond in the same way. In what ways can you help them learn from their reactions to this discovery? How can you capitalize on the diversity of their responses to teach tolerance and appreciation of different viewpoints?

Try your lesson ideas with your students and report back to your colleagues about what you learned.

Teaching Rounds

In the conclusion of Chapter One, "Lessons from a Coal Miner's Daughter," Teri Blanton identifies five lessons she has learned in becoming effective as a leader:

- Don't communicate from a place of anger.
- Reach people on the human level through stories.
- Foster dialogue instead of debate.
- Speak from the heart.
- Make ecological connections clear to others.

Talk with your colleagues about ways that you have taught one or more of these lessons in your classroom.

PART TWO: FROM ANGER TO ACTION IN OIL COUNTRY

Personal Reflection

Read Chapter Three, "The Heart of the Caribou." The chapter describes how indigenous leader Sarah James speaks for her people, the caribou, and the Earth when she educates others about oil drilling and how it adversely affects their traditional way of life and their environment. List some ways her perspective compares and contrasts with some of the perspectives that are predominant in American society.

Given the environmental crises we face, do we need to shift our society's priorities? Or do you think we can continue to live as we do and simultaneously heal the Earth?

If your answer to the first question is "Yes, we need to shift our priorities," create a list of the top five priorities that should guide our actions. If your answer is "Yes" to the second question, make a list of strategies that would allow us to continue to live as we do and, at the same time, work effectively to solve the major environmental issues of our times. Whether you answer "Yes" or "No" to these questions, explain your thinking. After recording your personal responses, discuss these questions and your answers with your colleagues.

Structured Conversation

It is difficult to find ways to engage students in real-world change. Featured in Chapter Four, the Rethinkers group is an excellent model of one way to do so through an after-school program. Brainstorm opportunities in your school community for involving students in taking direct action and influencing a local issue. For the first part of the conversation, suspend talking about the challenges that make this kind of learning activity difficult. After you have ten to fifteen ideas, identify those that have the most promise of success. This may vary by grade level. Now, choose someone to record the conversation and talk through the process and logistics of carrying out one or more of the ideas. (You may want to divide into smaller groups if there is interest in more than one possibility.) Finally, determine who is enthusiastic about pursuing this idea and establish a time to get together and move it forward.

Lesson Design

Read the sidebar "Using Circles to Cultivate Deep Listening" in Chapter Nine, which describes the Rethinkers process for working together. Join with your colleagues and develop lesson ideas that use circles such as those described in the sidebar to increase student participation and the quality of thought in class discussions.

- Think through the logistics of physically forming a circle. If your classroom does not have room, can you move to another space in the school?
- Develop and share individual ideas about rules for a circle discussion. For example, will you have different rules than those described in the sidebar? Will participants use a talking piece (an object such as a stick or a rock that is passed to each person when it is his or her time to speak)?
- What are some ways you will introduce circle discussions to your students?
- Consider how you will evaluate the "quality" of a discussion. How will you assess the quality and participation levels in circles, in comparison to those in regular class discussions?

Identify colleagues in the group who are willing to try conducting a circle with their students. Set aside time at a future meeting to share their experiences.

Teaching Rounds

Students portrayed in Chapter Four, "Beyond Whining," are solutions-oriented and, with the help of adults, find ways to take action in their community. With your colleagues, discuss ways you have moved students from a focus on a problem to a focus on problem-solving. What can students do to help solve a problem? If you have student work that conveys their thinking, bring it to share with your colleagues.

PART THREE: SHARED WATER: MOVING BEYOND BOUNDARIES

Personal Reflection

On the first page of Chapter Five, "Water Wars and Peace," Aaron Wolf talks about a "sudden jolt," which he defines as a "transformative shift in the room when suddenly everybody sees, understands, or experiences things differently from the way they did before."

With your colleagues, think back over your experiences as teachers and see if you can each identify one or more times when you experienced a sudden jolt with your students. Although every single student might have not made a transformative shift, try and recall times when a classroom reached a collective "aha!" moment. Share with your colleagues the conditions that led to those moments.

Structured Conversation

Read the Chapter Nine sidebar, "Transformation in the Classroom." Talk as a group about Wolf's "Four Worlds" frame of teaching and learning.

How can you incorporate Wolf's ideas into your classrooms? For example, how might you alter the physical conditions of space and time to improve teaching and learning? What strategies would enhance the emotional atmosphere in your classrooms? Are you interested in introducing a spiritual component to the atmosphere in your classrooms? (By "spiritual" we do not mean religious, but rather an inclusion of nonmaterial qualities such as compassion, responsibility, forgiveness, and interconnectedness.)

Lesson Design

As a group, check out Google Earth's virtual tour of oceans at http://earth. google.com/ocean/. Brainstorm ways you can use this website, or other maps and resources available from Google, to address one of the five practices of emotionally and socially engaged ecoliteracy: Understanding how nature sustains life.

GOING ON A VIRTUAL DIVE

"We cannot as a community conserve what we cannot see," Stanford University biologist Barbara Block has said.[1] To help us see some hidden parts of our planet, Google Earth now facilitates an exploration of the most expansive—and least explored—part of Earth: the oceans.

At the Google Earth website, http://earth.google.com/ocean/, students can take a virtual tour to learn about the joys, beauties, and mysteries of ocean life, from surfing hot spots, to the ocean sea floor, deep-sea vents, and the rich diversity of marine life. They can view videos of shipwrecks and follow, via satellite imagery, the predatory white shark on her undersea journey.

Students can also learn about what legendary National Geographic explorer Sylvia Earle calls "hope spots," areas of great biodiversity that are critical to the ocean's health. And most relevant to helping students cultivate an understanding of the interrelationship between human activity and nature, they can learn about how humans are affecting the ocean—a subject nearly as vast as the ocean itself.

Of the 139 million square miles of Earth's oceans, less than 4 percent remains unaffected by human activities, according to a study by Ben Halpern, a University of California marine ecologist whose work is featured on Google Earth. More than 40 percent of Earth's ocean area is heavily impacted by human activity.[2]

Google Earth showcases more than a dozen ways in which human activities affect the ocean—for example, through fishing, pollution, fertilizer runoff, acidification, commercial shipping, and climate change.[3]

The ocean feature of Google Earth was developed in collaboration with more than eighty organizations, including the National Geographic Society, the Smithsonian National Museum of History, and Woods Hole Oceanographic Institution.

Teaching Rounds

Read Chapter Six, "From Restoration to Resilience." This story describes the STRAW project (Students and Teachers Restoring a Watershed). Habitat restoration projects such as this one require that students practice perseverance and the ability to live with delayed gratification. These are practices that integrate emotional, social, and ecological intelligence. These practices may be unfamiliar to many of today's students, who are accustomed to instant rewards even for minimal efforts. Identify techniques that you have used or are considering to teach students to persevere, work harder, and accept delayed gratification for greater long-term returns.

PART FOUR: NOURISHING COMMUNITIES WITH FOOD

Personal Reflection

In Chapter Eight, "Forging the Food Justice Path," Tony Smith recognizes the need to improve school food as an equity issue. Do you think all students at your school have access to the same quality of food while at school? What about when they are out of school?

To answer the second question, draw a map of the school community, noting where students and their families acquire food. Do they seem to have equitable access to healthy, fresh food? Is there an overabundance of fast food establishments or liquor stores? Are there "food deserts" in the area? Examine your maps together and list any insights you gain from this exercise.

Structured Conversation

Discuss with colleagues the hidden messages that school food might send to students:

- Are the kinds of foods served at school in alignment with what students are taught about good nutrition?
- Does the atmosphere of the lunchroom (or wherever students eat) convey a sense of participating in an enjoyable act of eating together?
- Is the food affordable for most students? What about those for whom it is unaffordable? What happens when a student is too hungry to concentrate in class?
- Does the school provide healthy snacks during the state testing period but not at other times?
- Who makes and serves the school food? Do the students have a positive relationship with those people?

Lesson Design

Read the two Chapter Eight sidebars, "The Rise of School Food Reform" and "The Curriculum Connection." Refer to the academic standards adopted by your district and identify where these topics are addressed in the curriculum. Then share with your colleagues creative ways to teach students about food and nutrition.

Teaching Rounds

Food choices can be a sensitive subject in the classroom. Many children come from families where food is in limited supply, and their parents are under pressure to keep every family member's stomach full within a severely limited budget. Often, this means eating inexpensive and unhealthy fast foods, using food stamps, and making regular visits to a food bank or other hunger prevention program. In addition, an increasing number of students are obese and feel embarrassed when questioned about what they eat. There are also ethnic differences in family food preferences, and some kids are self-conscious that their family's food choices are different from typical American fare.

Discuss with your colleagues how you can involve students in learning about healthy food choices while still being sensitive to their cultural issues and other food-related issues.

These sample agendas apply the professional development guidelines described in Chapter Nine and the conversation strategies defined in Chapter Ten to the stories throughout this book. They outline sample formats for a one-hour learning circle and a half-day learning circle.

Sample Agenda for a One-Hour Learning Circle

Prior to the meeting, ask learning circle members to read Chapter Six, "From Restoration to Resilience." Set up the room so that everyone can sit in one large circle. Then follow these steps:

1. Conduct the welcome and opening ritual.

2. Review the group norms: Post and refer to them.

3. Discuss Chapter Six, "From Restoration to Resilience," using the Teaching Rounds strategy described on page 128.

4. Conduct the closing ritual and adjourn.

Sample Agenda for a Half-Day Professional Development

Set up the room so people can, at various times, sit in one large circle, sit in four smaller circles, work individually, and work in pairs. Then do the following:

1. Conduct the welcome and opening ritual.

2. Review the group norms: Post and refer to them.

3. Conduct a personal reflection exercise: Each educator reads over the five practices of ecoliteracy and makes notes about his or her individual level of understanding for each practice.

4. Gather in learning circles by grade level and share personal reflections. Then shift the focus to students. Discuss and record strategies to nurture the five practices of ecoliteracy at each grade level. Ask each grade-level group to determine its top three best ideas for nurturing ecoliteracy in the classroom.

5. Form a learning circle with the whole group and ask each grade level group to post its top ideas on the walls of the room. Talk about ways to strengthen the five practices of

ecoliteracy as a student moves up through the grades. Discuss ways to increase students' level of sophistication gradually so that a hypothetical school graduate would be ecoliterate (to an age-appropriate degree) upon leaving the school.

6. Distribute copies of Chapter One, "Lessons from a Coal Miner's Daughter," and draw attention to the two prompts for thinking about the story, which precede and follow each story. Give everyone twenty to thirty minutes to read the chapter. While your colleagues are reading, refer to the beginning of Chapter Ten and read the four professional development strategies for Part One, "Standing Strong on a Coal Mountain." Post them for the group to see.

7. When everyone has finished reading the chapter, ask each individual to select one of the four strategies and sit with others who chose the same strategy. Give the groups 30–45 minutes for discussion.

8. When the four groups have finished their conversations, ask the entire group to rejoin the large learning circle, then ask each group to summarize what it did.

9. Refer to the sidebar "How to Mine a Mountain" from Chapter One. Form pairs of teachers from the same grade level, and use large paper and colored pencils to create a map of the process of mountaintop mining as outlined in the sidebar. Ask each pair to discuss ways that mapping the mining process influenced their understanding and feelings about mining as an environmental, social justice, and health issue. Next ask the pairs to develop and try a mapping activity that they could use to help their students better grasp a different societal issue that would hold meaning for them. Post the second set of maps around the room. As an entire group, visit each map and listen to the originators describe their lesson ideas and what they think students would learn from them.

10. Ask everyone to return to a single learning circle. Invite them each to say a word or phrase that describes what they experienced during this professional development gathering.

11. Conduct the closing ritual and adjourn.

Conclusion

HANDS-ON HOPE

Our greatest hope in writing this book is that we have intrigued you with the idea that integrating emotional, social, and ecological intelligence lends the "secret sauce," if you will, to academic achievement and ecological well-being.

Socially and emotionally engaged ecoliteracy offers a wise and gentle process through which educators can positively join with their students to understand the reality of the ecological problems we now face—and creatively reimagine solutions. It allows school communities to transform topics that are too often presented as immobilizing or seemingly distant threats into highly relevant and worthy challenges that give young people the experience of working to make a difference in their particular part of the world.

Through socially and emotionally engaged ecoliteracy, educators can help create a safe container for exploration, effectively guide students in inquiries that reveal the connections that might otherwise remain unseen, and know when to push and when to pull back. Above all, it permits teachers to contribute to important, meaningful education that builds—in vitally important ways—on the social and emotional learning skills that decades of research have now shown foster student achievement, inspire better attitudes, and cultivate improved behavior. It also plants the seeds for a positive relationship with the natural world that can sustain a young person's interest and involvement for a lifetime.

As you travel this path, we hope you will take as guideposts the five practices of socially and emotionally engaged ecoliteracy:

- Developing empathy for all forms of life
- Embracing sustainability as a community practice
- Making the invisible visible
- Anticipating unintended consequences
- Understanding how nature sustains life

We also want to leave you with a reminder of the belief that sustained every leader we met as he or she confronted myriad challenges, from trying to stop mountaintop mining to remaking a food system. In the face of great challenges, you can never know whether you will succeed. As Wendell Berry has said, whether you succeed is not even the right question; what matters is that you try to do the right thing. Whether big or small, the act in itself counts.

Exactly how much those acts count can be thought of as our "handprint"—the positive alternative to our ecological footprint—or the sum total of all the things we do that benefit the planet. In fact, just as this book was going to press we learned of a new resource we'd like to recommend as a final offering. The website www.handprinter.org allows a student or group to calculate their handprint from all their ecologically beneficial activities. This includes not just their own activities, but also those of people they persuade. So if students go door-to-door with a tire gauge and pump, convincing neighbors to fill their tires to the proper level as a simple gas-saving and carbon-reducing move, that raises the handprint of those students, that class, and the entire community.

To embrace our relationship to the natural world fully is to know that although some of our actions have negative impacts, we also have boundless potential to create a positive effect. That effect might be helping students to reduce their school's dependence on oil, discover how their lives are connected to people who live near mountaintop removal sites, build resilience in a watershed, or increase equitable access to healthy food. Or, it might be what you do one day. In the end, it comes down to what caring educators do best: creating the conditions for learning that nurture hands-on hope.

INTRODUCTION

1. The idea of instability fomenting breakdown or breakthrough is based on what scholars have discovered about how systems change. As physicist Fritjof Capra has written:

   > Living systems generally remain in a stable state, even though energy and matter flow through them and their structures are continually changing. But every now and then such an open system will encounter a point of instability, where there is either a breakdown or, more frequently, a spontaneous emergence of new forms of order.

   > This spontaneous emergence of order at critical points of instability, which is often referred to simply as "emergence," is one of the hallmarks of life. It has been recognized as the dynamic origin of development, learning, and evolution. In other words, creativity—the generation of new forms—is a key property of all living systems. (http://www.ecoliteracy.org/essays/life-and-leadership-0)

2. According to the United Nations, the world's population is projected to grow to 9 billion by 2045, http://esa.un.org/unpd/wpp/index.htm.

3. In "Planetary Boundaries: Exploring the Safe Operating Space for Humanity" (*Ecology and Society* 14[2]: 32; http://www.ecologyandsociety.org/vol14/iss2/art32/), Johan Rockström, et al., discuss the nine planetary boundaries that cover the global biogeochemical cycles of nitrogen, phosphorus, carbon, and water; the major physical circulation systems of the planet (the climate, stratosphere, ocean systems); the biophysical features of Earth that contribute to the underlying resilience of its self-regulatory capacity (marine and terrestrial biodiversity, land systems); and two critical features associated with anthropogenic global change (aerosol loading and chemical pollution).

4. Fred Pearce, "From Ocean to Ozone: Earth's Nine Life-support Systems," *New Scientist*, February 24, 2010, http://www.newscientist.com/special/ocean-to-ozone-earths-nine-life-support-systems.

5. Rockström and others report that there are not yet enough data to assess the impact of chemical pollution and aerosol loading of the atmosphere.

6. Producing stainless steel requires a global supply chain involving more than 1,400 steps, each with its own environmental impact, according to Daniel Goleman and Gregory Norris, in "How Green is My Bottle," *New York Times*, April 19, 2009, http://www.nytimes.com/interactive/2009/04/19/opinion/20090419bottle.html.

7. "The world goes to town: After this year the majority of people will live in cities. Human history will ever more emphatically become urban history, says John Grimond," *The Economist*, http://www.economist.com/node/9070726.

8. The Center's initiatives include the following: Rethinking School Lunch, a planning framework based on a vision of healthy children who are ready to learn, "food literate" graduates, invigorated local communities, sustainable agriculture, and a healthy environment; the School Lunch Initiative with Alice Waters of the Chez Panisse Foundation and the Berkeley Unified School District; and Smart by Nature, a framework for schooling for sustainability based on work with schools and organizations in more than 400 communities across the United States and other countries.

9. "The Impact of Enhancing Students' Social and Emotional Learning: A Meta-Analysis of School-Based Universal Interventions," *Child Development*, January/February 2011, Volume 82, Number 1, pp. 405–32.

10. "Benefits of SEL," Collaborative for Academic, Social, and Emotional Learning, http://casel.org/why-it-matters/benefits-of-sel.

11. In 1998, for example, the State Education and Environment Roundtable (SEER), a cooperative endeavor of sixteen state departments of education, issued a report concluding that an environment-based context significantly improves student performance in reading, writing, math, science, and social studies, and enriches the overall school experience. A 2005 SEER report, "The Effects of Environment-Based Education on Student Achievement," also affirmed these findings, noting: "These engaging programs appear to better connect students to their learning by allowing them to take a more active role in their studies." See http://www.seer.org/pages/research/CSAPII2005.pdf

12. *Orion Magazine* webinar with Wendell Berry, Tim DeChristopher, and Teri Blanton, April 12, 2011.

CHAPTER ONE

1. For more information on mountaintop mining, see "Learn More about Mountaintop Removal Coal Mining," iLoveMountains.org, *http://www.ilovemountains.org/resources*.

2. Erin Brockovich is a former beauty contestant who helped successfully launch a case against Pacific Gas and Electric Company for the contamination of drinking water in the southern California town of Hinkley—despite the lack of any formal legal education. Actress Julia Roberts won an Oscar for her portrayal of Brockovich in a 2000 movie bearing her name.

3. A 50 percent increase in worldwide coal consumption is projected to occur from 2008 to 2035, according to the September 2011 U.S. Energy Administration report, "International Energy Outlook 2011," page 69, www.eia.gov/forecasts/ieo/coal.cfm.

4. National Energy Foundation, "Mining Coal," p. 1. http://www.coaleducation.org/lessons/twe/mcoal.pdf.

5. According to Margaret Palmer, director of the University of Maryland Center for Environmental Science, "Some three million pounds of explosives are detonated each day in West Virginia for coal mining, according to the U.S. Geological Survey, and the process shears up to 800 feet of elevation off each mountain peak." (From Sophia Yan, "In West Virginia, a Battle Over Mountaintop Mining," *Time*, March 12, 2010, http://www.time.com/time/health/article/0,8599,1971709,00.html.)

6. Brandon Keim, "Blowing the Top Off Mountaintop Mining," *Wired*, September 10, 2007, http://www.wired.com/science/planetearth/news/2007/09/mountaintop_mining.

7. Kennedy said, "If you came to the Hudson River and you tried to fill 25 feet of a Hudson River tributary, we would put you in jail, I guarantee it. If you tried to blow up a mountain in the Berkshires, the Adirondacks, or a mountain in Colorado, California or Utah, you would go to jail." See Jeff Goddell, "March to Stop Mountaintop Removal," *Rolling Stone*, June 16, 2011, http://www.rollingstone.com/politics/blogs/national-affairs/weeks-top-enviro-stories-zombie-nuclear-plant-lives-20110616.

8. Highlands Nature Center, "Biodiversity of the Southern Appalachians," http://www.wcu.edu/hbs/biodiversity.htm.

9. Center for Health and the Global Environment, Harvard Medical School, "Mining Coal, Mounting Costs: The life cycle consequences of coal," http://chge.med.harvard.edu/resource/mining-coal-mounting-costs-life-cycle-consequences-coal.

10. These disparities, the researchers noted, cannot be accounted for by other factors such as smoking, diet, income, or access to health care. (M. A. Palmer, et al., "Mountaintop Mining Consequences," *Science,* January 8, 2010, http://www.sciencemag.org/content/327/5962/148.summary.)

11. Dan Barry, "As the Mountaintops Fall, a Coal Town Vanishes," *The New York Times,* April 12, 2011, http://www.nytimes.com/2011/04/13/us/13lindytown.html?_r=1&scp=1&sq=lindytown&st=cse.

12. Center for Climate and Energy Solutions, "Coal and Climate Change Facts," http://www.pewclimate.org/global-warming-basics/coalfacts.cfm.

13. Kristin Johannsen, et al., eds, *Missing Mountains: We went to the mountaintop but it wasn't there* (Wind Publications, 2005).

14. *Ibid.*

15. Wendell Berry, *Life is a Miracle: An Essay Against Modern Superstition* (Counterpoint, 2000).

16. Wendell Berry, "Speech Against the State Government, Frankfort, 2/14/08," Kentuckians for the Commonwealth, http://www.kftc.org/our-work/general-assembly/stream-saver-bill/Wendell%20Berry%202-14–08.pdf.

17. Brett Barrouquere, "Kentucky, coal group sue EPA over permits," *The Register-Herald,* http://www.register-herald.com/todaysfrontpage/x693282295/Kentucky-coal-group-sue-EPA-over-permits.

18. Paul R. Epstein, et al., "Full cost accounting for the life cycle of coal," *Ecological Economics Reviews*, February 2011, pp. 73–98, http://solar.gwu.edu/index_files/Resources_files/epstein_full%20cost%20of%20coal.pdf.

19. Erik Hungerbuhler, "KFTC members host Governor Beshear on a tour of Eastern Kentucky," Kentuckians for the Commonwealth, http://www.kftc.org/blog/archive/2011/05/04/archive/2011/04/08/kftc-members-host-governor-beshear-on-a-tour-of-eastern-kentucky.

20. Erik Reese, *Lost Mountain: A year in the Vanishing Wilderness* (Riverhead Books, 2006).

21. *Ibid.*

22. *Ibid.*

CHAPTER TWO

1. Sue Sturgis, "Latest Duke coal plant challenge targets Appalachian mountaintop removal," Facing South, The Institute for Southern Studies, http://southernstudies.org/2008/03/latest-duke-coal-plant-challenge-targets-appalachian-mountaintop-removal.html.

2. Wendell Berry, "Wendell Berry's NKU Commencement Address Spring 2009 – Part 2" [Video] Retrieved August 24, 2011, from http://www.youtube.com/watch?v=oRgbLJnjwsQ.

3. Patty Tarquino left Kentucky after nearly seven years of working for Kentuckians for the Commonwealth because of concerns about the impact of the work on her own health. During one year she lived there, she said, her drinking water was contaminated with diesel fuel on three separate occasions.

4. From the Nature Conservancy Bad Branch State Nature Preserve website at http://www.nature.org/ourinitiatives/regions/northamerica/unitedstates/kentucky/placesweprotect/bad-branch-state-nature-preserve.xml.

5. The Gilberts eventually proved successful in their fight, and the coal company withdrew its permit application to build a catch pond, according to Melissa Duley, in "Paradise Purloined," *Louisville Magazine*, September 2007, http://www.loumag.com/articledisplay.aspx?id=24449917.

6. John McQuaid, "Mountaintop Mining Legacy: Destroying Appalachian Streams," *Yale Environment 360,* July 20, 2009, http://e360.yale.edu/content/feature.msp?id=2172.

7. This contaminated water raised the level of lead and thallium (which cause birth defects and other disorders) in nearby water sources, according to Shaila Dewan, in "Tennessee Ash Flood Larger than Initial Estimate," *New York Times,* December 26, 2008, http://www.nytimes.com/2008/12/27/us/27sludge.html.

8. The students visited B&W Resources, Inc., in Manchester, Kentucky.

9. Steele had been strongly influenced by a study abroad experience in Thailand conducted through the Council of International Educational Exchange. He currently serves on the steering committee of the Education Network for Global and Grassroots Exchange, a network of students, educators, and community organizers who work to transform learning experiences into lifelong connections and cooperative action between peoples and social movements working for a just and sustainable world.

10. As a result of Milliken's prompting and support, ninety-six geothermal wells were dug in Spartanburg, which Dorrance says now provide an ecological and cost-effective energy source.

CHAPTER THREE

1. U.S. Fish & Wildlife Service, "Frequently Asked Questions about Caribou," http://arctic.fws.gov/carcon.htm.
2. The only tribe that lives farther north is the Inuit.
3. Peter Matthiessen, "In the Great Country," in Subhankar Banerjee, *Arctic National Wildlife Refuge: Seasons of Life and Land,* p. 44 (Mountaineers Books, 2003).
4. *Ibid.,* p. 40
5. James shared the 2002 Goldman Environmental Prize with two other Gwich'in elders, the late Jonathon Solomon from Fort Yukon, Alaska, and Norma Kassi from Old Crow, Canada.
6. Mark Guarino, "Alaska oil pipeline has safety and environmental risks," *The Christian Science Monitor,* February 12, 2011, http://www.csmonitor.com/Environment/2011/0212/Alaska-oil-pipeline-has-safety-and-environmental-risks.
7. "National Petroleum Reserve-Alaska," *Wikipedia,* http://en.wikipedia.org/wiki/National_Petroleum_Reserve–Alaska.
8. Becky Bohrer, "Obama administration signals interest in more Alaska oil," Associated Press, August 8, 2011, http://www.cbsnews.com/stories/2011/08/08/national/main20089836.shtml.
9. U.S. Energy Information Administration, http://www.eia.gov/countries/index.cfm#countrylist.
10. Eric Fox, "Bubbling crude: America's top 6 oil-producing states," msnbc.com, http://www.msnbc.msn.com/id/43085246/ns/business-oil_and_energy/t/bubbling-crude-americas-top-oil-producing-states/.
11. Caribou are of the same species as reindeer, a domesticated variety of caribou used primarily for pulling sleds in Scandinavia and Serbia. (U.S. Fish & Wildlife Service, "Frequently Asked Questions about Caribou," http://arctic.fws.gov/carcon.htm).
12. Peter Matthiessen, "Inside the Endangered Arctic Refuge," *The New York Review of Books,* October 19, 2006.
13. Even today, the Gwich'in refer to themselves as having, at least figuratively, a heart shared with the caribou. "We don't mean physically," said James, "but

that we have a heart for this, for the caribou. They are in our heart because we are protecting them. Of course, they probably have a heart for us, too, because we're protecting them."

14. Alaska Community Database Community Information Summaries (CIS), http://www.commerce.state.ak.us/dca/commdb/CIS.cfm?Comm_Boro_Name=Arctic%20Village.

15. Rick Bass, *Caribou Rising: Defending the Porcupine Herd, Gwich-'in Culture, and the Arctic National Wildlife Refuge* (Sierra Club Books, 2004).

16. Alaska Community Database Community Information Summaries (CIS).

17. 2006–2010 American Community Survey (ACS), cited in Alaska Community Database Community Information Summaries (CIS), http://www.commerce.state.ak.us/dca/commdb/CIS.cfm?Comm_Boro_Name=Arctic%20Village.

18. Matthiessen, "In the Great Country," p. 41.

19. Elizabeth Shogren, "For 30 Years, a Political Battle over Oil and ANWR," National Public Radio, November 10, 2005, http://www.npr.org/templates/story/story.php?storyId=5007819.

20. *Ibid.*

21. Gwich'in Steering Committee, Gwich'in Niintsyaa (Resolution), "Resolution to Prohibit Development in the Calving and Post-Calving Grounds of the Porcupine caribou herd," http://www.gwichinsteeringcommittee.org/gwich-inniintsyaa.html.

22. For more information on the Alaskan oil spill, see Pamela A. Miller, "Exxon Valdez Oil Spill: Ten Years Later. Technical Background Paper for Alaska Wilderness League," *Arctic Connections* 3/99, http://arcticcircle.uconn.edu/SEEJ/Alaska/miller2.htm. Also see "A Brief History of the Exxon Valdez Disaster," http://www.time.com/time/photogallery/0,29307,1887217,00.html.

23. Hazardous Materials Response and Assessment Division, *Oil Spill Case Histories 1967–1991, Report No. HMRAD 92–11*, September 1992. Seattle: National Oceanic and Atmospheric Administration, p. 80. Retrieved March 10, 2008 (PDF).

24. Mary Pemberto, "Arctic National Wildlife Refuge: Obama Urged to Give National Monument Status to Protect from Oil Drilling," *The Huffington Post,* December 4, 2010, http://www.huffingtonpost.com/2010/12/06/arctic-national-wildlife-_n_792531.html.

25. U.S. Energy Information Administration, "2009 World Oil Consumption," http://www.eia.gov/countries/index.cfm?view=consumption.

CHAPTER FOUR

1. Rethink, http://www.therethinkers.com/what-weve-done-2/.

2. Lisa Bennett, "Now for the Kids' Say on the Spill," *The Huffington Post*, July 14, 2010, http://www.huffingtonpost.com/lisa-bennett/now-for-the-kids-say-on-t_b_644766.html. "New Orleans Students Challenge Schools on Use of Oil," *The Huffington Post*, July 22, 2010, http://www.huffingtonpost.com/lisa-bennett/new-orleans-students-chal_b_656128.html.

3. Jane Wholey, Social Entrepreneurs of New Orleans, http://seno-nola.org/index.php?option=com_content&view=article&id=48&Itemid=46.

4. "FAQs, Hurricane Katrina Relief," http://www.hurricanekatrinarelief.com/faqs.html#Why%20did%20the%20levees%20break%20in%20New%20Orleans. For more information about the hurricane, see Richard D. Knabb, et al., "Tropical Cyclone Report: Hurricane Katrina: 23–30 August 2005," August 2006, http://www.nhc.noaa.gov/pdf/TCR-AL122005_Katrina.pdf.

5. Wholey, Social Entrepreneurs of New Orleans.

6. Loreal Lynch, "Achieving Greater Impact Through Model Sharing," *Stanford Social Innovation Review,* May 4, 2010, http://www.ssireview.org/opinion/entry/achieving_greater_impact_through_model-sharing/.

7. Sarah Laskow, "Necessity Is the Mother of Invention," *The Daily Beast,* http://www.thedailybeast.com/newsweek/2010/08/26/new-orleans-s-charter-school-revolution.html.

8. Amanda M. Fairbanks, "Fixing the Broken Parts: Can Schools Save New Orleans?" *GOOD Cities*, July 15, 2010, http://www.good.is/post/fixing-the-broken-parts-can-schools-save-new-orleans/.

9. Elizabeth R. Hinde, "School Culture and Change: An Examination of the Effects of School Culture on the Process of Change," *Essays in Education,* 12 (winter: 2004), www.usca.edu/essays/vol122004/hinde.pdf.

10. Survey results were analyzed in consultation with Ted Quant, director of Loyola University's Twomey Center for Peace Through Justice. (Daniel Goleman, et al., "Forging New Norms in New Orleans" *Teacher Education Quarterly*, Fall 2010)

11. Campbell Robertson, "Gulf of Mexico Has Long Been Dumping Site," *New York Times*, July 30, 2010, http://www.nytimes.com/2010/07/30/us/30gulf.html?pagewanted=all.

12. *Ibid.*

13. Seth Broenstein, "Petroleum-based products are around us, in us," Associated Press, June 12, 2010, http://www.postandcourier.com/news/2010/jun/12/petroleum-based-products-are-around-us-in-us/.

14. U.S. Energy Information Administration, "Oil (petroleum) Basics," http://www.eia.gov/kids/energy.cfm?page=oil_home-basics.

15. Hydro, nuclear, and other (geothermal, solar, wind, and wood and waste) electric power generation ranked fourth, fifth, and sixth, respectively, as primary energy sources in 2006, accounting for 6.3, 5.9, and 1.0 percent, respectively, of world primary energy production. (U.S. Energy Information Administration, http://www.eia.gov/iea/overview.html)

16. The manager appointed to oversee the final well tests had so little experience that, in his own words, he was on the rig to "learn about deep water." (Ben Casselman and Russell Gold, "BP Decisions Made Well Vulnerable," *The Wall Street Journal,* May 27, 2010, http://online.wsj.com/article/SB10001424052748704026204575266560930780190.html)

17. Even a catastrophic offshore spill, regulators said, should never reach shore. (Neil King Jr. and Keith Johnson, "BP Relied on Faulty US Data," *The Wall Street Journal,* June 24, 2010, http://online.wsj.com/article/SB10001424052748703900004575325131111637728.html)

18. Lisa Bennett, "New Orleans Students Challenge Schools on Use of Oil," *The Huffington Post,* July 22, 2010, http://www.huffingtonpost.com/lisa-bennett/new-orleans-students-chal_b_656128.html.

CHAPTER FIVE

1. "Talking Point: Ask Boutros Boutros Ghali," BBC News, June 10, 2003, http://news.bbc.co.uk/2/hi/talking_point/2951028.stm.

2. "Peace in the pipeline," BBC News, February 13, 2009, http://news.bbc.co.uk/2/hi/science/nature/7886646.stm.

3. "The Freshwater Guide," Coordinating Committee for International Voluntary Service, p. 11, http://www.unesco.org/ccivs/New-SiteCCSVI/CcivsOther/Documents/FreshwaterGuideEN.pdf.

4. Maywa Montenegro, "The Truth About Water Wars," *SEED,* May 14, 2009, http://seedmagazine.com/content/article/the_truth_about_water_wars.

5. Kevin Watkins and Anders Berntell, "A global problem: How to avoid war over water," *The New York Times,* August 23, 2006, http://www.nytimes.com/2006/08/23/opinion/23iht-edwatkins.2570814.html?pagewanted=all.

6. Books include *Water Wars: Privatization, Pollution, and Profit,* by Vandana Shiva (South End Press, 2002); *Water Wars: Drought, Flood, Folly and the Politics of Thirst,* by Diane Raines Ward (Riverhead Books, 2002); and *The Water Wars,* by Cameron Stracher (Sourcebooks Fire, 2011).

7. *Blue Gold: World Water Wars*, a 90-minute documentary by Sam Bozzo, was released in 2008, http://www.bluegold-worldwaterwars.com/press_kit/blue_gold_press_kit.pdf.

8. Charles Fishman, *The Big Thirst: The Secret Life and Turbulent Future of Water* (Free Press, 2011).

9. Fishman, p. 2.

10. *Ibid.*, p. 4.

11. United States Geological Survey, "Thermoelectric Power Water Use," http://ga.water.usgs.gov/edu/wupt.html.

12. United States Geological Survey, "The Water Cycle: Evapotranspiration," http://ga.water.usgs.gov/edu/watercycleevapotranspiration.html.

13. United States Geological Survey, "Irrigation Water Use," http://ga.water.usgs.gov/edu/wuir.html.

14. United States Geological Survey, "Public-Supply Water Use," http://ga.water.usgs.gov/edu/wups.html.

15. United States Geological Survey, "Total Water Use in the United States, 2005," http://ga.water.usgs.gov/edu/wateruse-total.html.

16. The growth of water privatization since the 1990s, the inefficiencies of faulty and aging infrastructure, and the pollution of fresh water drinking supplies are other factors that fuel concerns about water scarcity.

17. Kathy Jesperson, "What's the Word on Water at the USDA?" National Drinking Water Clearinghouse, West Virginia University, http://www.nesc.wvu.edu/ndwc/pdf/OT/OTs96.pdf.

18. "Tapping the oceans," *The Economist*, June 5, 2008, http://www.economist.com/node/11484059.

19. *Ibid.*

20. Karen E. Lange, "Get the Salt Out," *National Geographic*, March 15, 2010, http://ngm.nationalgeographic.com/big-idea/09/desalination.

21. Kathryn Kranhold, "Water, Water, Everywhere," *The Wall Street Journal*, January 17, 2008, http://online.wsj.com/article/SB120053698876396483.html.

22. "Tapping the Oceans," *The Economist*.

23. Michael Wines, "China Takes a Loss to Get Ahead in the Business of Fresh Water," *New York Times*, Oct. 25, 2011, http://www.nytimes.com/2011/10/26/world/asia/china-takes-loss-to-get-ahead-in-desalination-industry.html?pagewanted=all.

24. Joey Peters and Climate Wire, "Climate Change Could Leave 1 Billion Urbanites High and Dry by 2050," *Scientific American*, April 4, 2011, http://www.scientificamerican.com/article.cfm?id=climate-change-one-billion-urbanites-water-shortage.

25. Alex Prud'homme, *The Ripple Effect: The Fate of Freshwater in the Twenty-first Century* (Scribner, 2011).

26. David Molden, ed., *Water for Food, Water for Life: A Comprehensive Assessment of Water Management in Agriculture,* International Water Management Institute, 2007, http://www.iwmi.cgiar.org/assessment/files_new/synthesis/Summary_SynthesisBook.pdf.

27. The United Nations, "Water Scarcity," http://www.un.org/waterforlifedecade/scarcity.shtml.

28. "Industrial Water Use," Water Science for Schools, U.S. Geological Survey, U.S. Department of the Interior, http://ga.water.usgs.gov/edu/wuin.html.

29. "Glaciers Are Melting Faster Than Expected, UN Reports," *Science Daily,* March 17, 2008, http://www.sciencedaily.com/releases/2008/03/080317154235.htm.

30. United Nations, International Decade for Action: Water for Life, 2005–2015, http://www.un.org/waterforlifedecade/scarcity.shtml.

31. Aaron Wolf is the nephew of Daniel Goleman, one of the authors of this book.

32. Kevin Watkins and Anders Berntell, "A Global Problem: How to Avoid War over Water—Editorials & Commentary—International Herald Tribune," *New York Times,* August 23, 2006, http://www.nytimes.com/2006/08/23/opinion/23iht-edwatkins.2570814.html?pagewanted=all.

33. Richard Cowen, "Middle East Water: The Geopolitics of Middle Eastern Water," University of California, Davis, Geology Department, http://mygeologypage.ucdavis.edu/cowen/~GEL115/115CHXXMideastwater.html.

34. Don Belt, "Parting the Waters," *National Geographic,* April 2010, p. 158.

35. Sandra L. Postel and Aaron T. Wolf, "Dehydrating Conflict," *Foreign Policy,* September 18, 2001, http://www.foreignpolicy.com/articles/2001/09/01/dehydrating_conflict.

36. Stan Mack, "How Is Water Cleaned Naturally?" eHow, http://www.ehow.com/how-does_5163235_water-cleaned-naturally.html.

37. The SODIS method was developed by Eawag, the Swiss Federal Institute of Aquatic Science and Technology (http://www.sodis.ch/about/eawag/index_EN).

38. Mark D. Sobsey, "Managing Water in the Home: Accelerated Health Gains from Improved Water Supply," World Health Organization, http://www.emro.who.int/ceha/pdf/Doc-managing.pdf.

39. Navigating Peace Initiative, "Water Conflict and Cooperation: Looking Over the Horizon," Woodrow Wilson International Center for Scholars, http://www.wilsoncenter.org/sites/default/files/ECSPReport13_NavigatingPeace.pdf.

40. Wolf identifies the remaining ninety-six conflicts as resolved neutrally. (Aaron Wolf, et al., "International Waters: Identifying Basins at Risk," 2003, *Water Policy* 5(1), 31–62. Cited in Aaron T. Wolf, et. al, "Water Can Be a Pathway to Peace, Not War," http://www.wilsoncenter.org/sites/default/files/ECSPReport13_NavigatingPeace.pdf.)

41. One reason hostilities seem more prevalent than cooperation, Wolf suggests, is that hostilities make news and agreements do not. In Oregon, for example, people tend to be aware of the controversy over water use in the Klamath River Basin because irrigation for agriculture was temporarily halted in 2001 to protect endangered salmon and lake fish during a severe drought. "But no one knows about the Walla Walla, the Umatilla, or the Grande Ronde," he adds, "where tribes, ranchers, environmentalists and urbanites have all been collaborating quietly to work proactively to manage regional water stresses." (Aaron Wolf, personal correspondence, Oct. 2, 2011)

42. Wendy Barnaby, "Do Nations Go to War Over Water?" *Nature*, March 2009, http://www.nature.com/nature/journal/v458/n7236/full/458282a.html.

43. Aaron T. Wolf, "Healing the Enlightenment Rift: Rationality, Spirituality and Shared Waters," *Journal of International Affairs*, Spring/Summer 2008, Vol. 61, No. 2., p 60.

44. Vahid Alavian was a World Bank water advisor at the time of this conversation. He retired in October 2011.

45. Email correspondence with Vahid Alavian, November 2, 2011.

46. Wolf, "Healing the Enlightenment Rift," p. 63.

47. *Ibid.*, p. 55.

48. Huston Smith, *Forgotten Truth: The Common Vision of the World's Religions,* (HarperOne, 1992), cited in "Healing the Enlightenment Rift," p. 63.

49. Abraham Maslow's hierarchy of needs include physiological needs, security needs, social needs, esteem needs, and self-actualizing needs.

50. Wolf, "Healing the Enlightenment Rift," p. 63.

51. *Ibid.*, p. 63.

52. *Sharing Water, Sharing Benefits: Working Towards Effective Transboundary Water Resources Management* is available for download at http://www.scribd.com/doc/52803099/Sharing-Water-Sharing-Benefits-Working-Towards-Effective-Trans-Boundary-Water-Resources-Management-9789231041679.

53. Jerome Delli Priscoli and Aaron T. Wolf, *Managing and Transforming Water Conflicts* (Cambridge Press, 2009).

CHAPTER SIX

1. Temple Grandin, "Preventing Bull Accidents," Colorado State University, updated June 2006, http://www.grandin.com/behaviour/principles/preventing.bull.accidents.html.

2. PRBO Conservation Science, "STRAW Project," http://www.prbo.org/cms/192.

3. David W. Orr, *Ecological Literacy: Education and the Transition to a Postmodern World* (State University of New York Press, 1991).

4. Thanks to Michael K. Stone, senior editor, Center for Ecoliteracy. This section draws heavily on his article, "Solving for Pattern: The Straw Project, published in *Whole Earth Magazine,* Spring 2001.

5. The California State Adopt-A-Species Program is now defunct.

6. Douglas H. Chadwick, "Silent Streams," *National Geographic,* April 2010, p. 118, http://ngm.nationalgeographic.com/2010/04/freshwater-species/chadwick-text/1.

7. *Ibid.*, p. 119.

8. Meanwhile, the number of waterways channeled to allow for ease of navigation grew from fewer than 9,000 in 1900 to almost 500,000 in 1999. (World Resources Institute, "The decline of freshwater ecosystems," http://www.wri.org/publication/content/8391)

9. The Nature Conservancy, "Rivers and Lakes: Reducing the Ecological Impact of Dams," http://www.nature.org/ourinitiatives/habitats/riverslakes/reducing-the-ecological-impact-of-dams.xml.

10. U.S. Environmental Protection Agency, "EPA's Healthy Watersheds Initiative," http://water.epa.gov/polwaste/nps/watershed/upload/2009_08_05_NPS_healthywatersheds_highquality_hwi.pdf.

11. David Biello, "Oceanic Dead Zones Continue to Spread," *Scientific American,* August 15, 2008, http://www.scientificamerican.com/article.cfm?id=oceanic-dead-zones-spread. See also Bina Venkataraman, "Ocean 'Dead Zones' on the Rise," *New York Times,* August 15, 2008, http://www.nytimes.com/2008/08/14/health/14iht-15oceans.15304270.html.

12. Roger Highfield, "Ocean dead zones free of oxygen double every decade," *The Telegraph,* August 14, 2008, http://www.telegraph.co.uk/science/science-news/3349611/Ocean-dead-zones-free-of-oxygen-double-every-decade.html.

13. Virginia Institute of Marine Science, "Study Shows Continued Spread of 'Dead Zones'; Lack Of Oxygen Now a Key Stressor on Marine Ecosystems," *Science Daily,*

August 14, 2008, http://www.sciencedaily.com/releases/2008/08/080814154325. htm.

14. Biello, "Oceanic Dead Zones Continue to Spread."

15. Robert J. Diaz and Rutger Rosenberg, "Spreading Dead Zones and Consequences for Marine Ecosystems," *Science*, August 15, 2008: Vol. 321 no. 5891 pp. 926–29, http://www.sciencemag.org/content/321/5891/926.abstract?ijkey=Lk77EfI/7o LkY&keytype=ref&siteid=sci.

16. "Dead water: Too much nitrogen being washed into the sea is causing dead zones to spread alarmingly," *The Economist,* May 15, 2008, http://www.economist.com/node/11367884?story_id=11367884.

17. According to the National Oceanic and Atmospheric Administration the term "dead zone" is a more common term for hypoxia, which refers to a reduced level of oxygen in the water (http://oceanservice.noaa.gov/facts/deadzone.html).

18. Johan Rockström, et al., "Planetary Boundaries: Exploring the Safe Operating Space for Humanity," *Ecology and Society* 14(2):32, http://www.ecologyand-society.org/vol14/iss2/art32/.

19. The bacteria essential to the nitrogen cycle help to balance the amount of nitrogen in the soil and in the atmosphere. However, since farming depletes soil over time, farmers add fertilizers to replenish nitrogen and other nutrients and to increase plant growth.

20. Biello, "Oceanic Dead Zones Continue to Spread."

21. Marin County Watershed Program, "Stemple Creek Watershed," http://www.marinwatersheds.org/stemple_creek.html.

22. American Society of Landscape Architects, "The Brave New World of Ecological Restoration," *The Dirt,* April 6, 2011, http://dirt.asla.org/2011/04/06/the-brave-new-world-of-ecological-restoration/.

23. Wikipedia, "Stream Restoration," http://en.wikipedia.org/wiki/Stream_restoration#cite_note-5.

24. Brendan Borrell, "Real Good or Feel-Good? Does Ecosystem Restoration Pay Off?" *Scientific American,* July 30, 2009, http://www.scientificamerican.com/article.cfm?id=does-ecosystem-restoration-pay-off.

25. José M. Rey Benayas, et al., "Enhancement of Biodiversity and Ecosystem Services by Ecological Restoration: A Meta-Analysis," *Science*, July 2009.

26. Independent scientists estimate the number of people affected by perchlorate at 20 to 40 million. (Juliet Eilperin, "EPA Unlikely to Limit Perchlorate in Tap Water," *Washington Post,* September 22, 2008, http://www.washingtonpost.com/wp-dyn/content/article/2008/09/21/AR2008092102352.html)

27. Environmental Working Group, "EWG National Drinking Water Database," http://www.ewg.org/tap-water/executive-summary.

28. Charles Duhigg, "U.S. Bolsters Chemical Restrictions for Water," *New York Times,* March 22, 2010, http://www.nytimes.com/2010/03/23/business/23water.html.

29. CNN Wire Staff, "EPA to set limits on chemicals in drinking water," CNN, February 2, 2011, http://www.cnn.com/2011/HEALTH/02/02/epa.water.chemical/index.html.

CHAPTER SEVEN

1. To learn about La Semilla Food Center, visit the website at http://www.lasemillafoodcenter.org/.

2. The Youth Food Policy Council is conducted in collaboration with the Colonias Development Council, an organization dedicated to improving the quality of life in communities along the United States–Mexico border. See http://www.lasemillafoodcenter.org/programs.html for more information.

3. According to the New Mexico Collaboration to End Hunger, in 2010, New Mexico was ranked the twelfth most food insecure state. See http://www.endnmhunger.org/about.html for more information.

4. Trust for America's Health, "New Report: New Mexico is 33rd Most Obese State in the Nation," July 7, 2011, http://healthyamericans.org/reports/obesity2011/release.php?stateid=NM.

5. According to the U.S.–Mexico Border Diabetes Prevention and Control Project. See also http://healthyamericans.org/reports/obesity2011/release.php?stateid=NM.

6. Roni Neff, "Food Matters: How What We Eat Affects Our Health and the Planet," *Imagine,* January/February 2009, http://www.jhsph.edu/bin/s/a/food-matters.pdf.

7. Norman Borlaug was an American agronomist, humanitarian, and Nobel laureate who has been called "the father of the Green Revolution" (http://en.wikipedia.org/wiki/Norman_Borlaug).

8. In 2011, The Berry Center was established at New Castle, Kentucky. Its mission is "to continue his work by bringing focus, knowledge, and cohesiveness to the work of changing our ruinous industrial agriculture system into a culture that uses nature as the standard, that accepts no permanent damage to the ecosphere, and that takes into consideration human health in local communities." (http://www.berrycenter.org/mission-and-vision/)

9. Tara Lohan, ed., *Water Matters* (AlterNet Books, 2010), cited in Wenonah Hauter, "Industrial Agriculture's Water Use: It's Time for Change," *Mother Earth News*, June 17, 2011, http://www.motherearthnews.com/sustainable-farming/industrial-agriculture-water-use-ze0z11zkon.aspx.

10. American Public Health Association, "Toward a Healthy, Sustainable Food System," November 6, 2007, http://www.apha.org/advocacy/policy/policysearch/default.htm?id=1361.

11. UN News Service, "UN expert makes case for ecological farming practices to boost food production," UN News Centre, March 8, 2011, http://www.un.org/apps/news/story.asp?NewsID=37704&Cr=farming. To read De Schutter's report, visit http://www.srfood.org/index.php/en/component/content/article/1174-report-agroecology-and-the-right-to-food.

12. "Colonia communities" are rural communities along the United States-Mexico border that may lack some of the most basic living necessities, such as potable water and sewer systems, electricity, paved roads, and safe and sanitary housing (http://www.hud.gov/groups/farmwkercolonia.cfm).

13. David Martin, "All McNuggets Not Created Equal," CNN, June 25, 2010, http://thechart.blogs.cnn.com/2010/06/25/a-tale-of-2-nuggets/.

CHAPTER EIGHT

1. Matt Byers, et al., "Life and Death from Unnatural Causes: Health and Social Inequity in Alameda County," Alameda County Public Health Department, August 2008, http://www.acphd.org/media/53628/unnatcs2008.pdf.

2. Smith previously served as deputy superintendent of instruction, innovation, and social justice for the San Francisco Unified School District, and prior to that he was superintendent of the Emeryville Unified School District in Emeryville, California.

3. "Jamie Oliver's Food Revolution" television program is no longer on the air, but his website offers news, recipes, and information at http://www.jamieoliver.com/us/foundation/jamies-food-revolution/school-food.

4. E. A. Finkelstein, et, al., "Annual Medical Spending Attributable to Obesity: Payer- and Service-Specific Estimates," *Health Affairs*, 2009, 28(5), pp. 822–31.

5. Kenneth E. Thorpe, Ph.D., "The Future Costs of Obesity: National and State Estimates of the Impact of Obesity on Direct Health Care Expenses," www.nccor.org/downloads/CostofObesityReport-FINAL.pdf.

6. Nicole Larson and Mary Story, "Are 'Competitive Foods' Sold at School Making Our Children Fat?" *Health Affairs*, 2010, 29(3), pp. 430–35.

7. Mary Story, "The Third School Nutrition Dietary Assessment Study: Findings and Policy Implications for Improving the Health of US Children," *Journal of the American Dietetic Association*, 2009, 109(2), S7-S13.

8. Action for Healthy Kids, *The Learning Connection: The Value of Improving Nutrition and Physical Activity in Our Schools*, October 2004, http://www.actionforhealthykids.org/for-schools/resources/research-and-reports/the-learning-connection-the-value-of-improving-nutrition-and-physical-activity-in-our-schools.html.

    K. Alaimo, et al., "Food insufficiency and American school-aged children's cognitive, academic and psychosocial development," *Pediatrics*, 2001, 108(1), pp. 44–53.

    J. M. Murphy, "Breakfast and Learning: An Updated Review," *Current Nutrition & Food Science*, 2007, 3, pp. 3–36.

    School of Nutrition Science and Policy, Tufts University, *The Link Between Nutrition and Cognitive Development in Children*, 1998. For excerpts from this article, see http://www.eecom.net/mfsp/projects_school_links.pdf.

    Marco Visscher, "You do what you eat," *Ode Magazine*, September 2005, http://www.odemagazine.com/doc/26/you_do_what_you_eat/.

9. Oakland Unified School District has a 37 percent dropout rate. The national average is 25 percent. For more information, see Gretchen Weber, "New High School Dropout Data: Oakland at 37 Percent," KQED News, August 11, 2011, http://blogs.kqed.org/newsfix/2011/08/11/new-high-school-dropout-data-oakland-at-37-percent/ and Mary Bruce, "25% of Students Fail to Graduate on Time, Report Says," June 2, 2010, ABC News, http://abcnews.go.com/blogs/politics/2010/06/25-of-students-fail-to-graduate-on-time-report-says/.

10. Geoffrey Canada leads the Harlem Children's Zone, a hugely ambitious network of charter schools renowned for its cradle-to-college approach. "Geoffrey Canada," *New York Times*, December 10, 2010, http://topics.nytimes.com/topics/reference/timestopics/people/c/geoffrey_canada/index.html.

11. Jabari Mahiri and Derek Van Rheenen, *Out of Bounds: When Scholarship Athletes Become Academic Scholars* (Peter Lang Publishing, 2010).

12. Douglas A. McIntyre, et al., "The Most Dangerous Cities in America," Yahoo! Finance, May 25, 2011, http://finance.yahoo.com/news/pf_article_112804.html.

13. Shoshana Walter, "Shootings Soar in Oakland; Children Often the Victims, *The New York Times*, January 7, 2012, http://www.nytimes.com/2012/01/08/us/children-are-often-victims-as-shootings-soar-in-oakland.html.

14. "Mayor Quan unveils 2011–2013 budget proposal," *Oakland Local*, April 30, 2011, http://oaklandlocal.com/article/mayor-quan-unveils-2011–2013-budget-proposal.

15. "The 45 Top Places to Go in 2012," *New York Times*, January 6, 2012, http://travel.nytimes.com/2012/01/08/travel/45-places-to-go-in-2012.html?pagewanted=all.

16. In the Oakland flatlands, where the median household income is $32,000, there's an average of one supermarket per 93,126 residents, according to a 2009 report by the Hope Collaborative. The same report found that in the Oakland Hills, where the median household income is more than $58,000, there's an average of one supermarket per 13,778 residents. (Angela Bass, et al., "Oakland's food divide: No grocery store in sight," OaklandNorth, http://oaklandnorth.net/few-food-choices/)

17. Phillip R. Kaufman, et al., "Do the Poor Pay More for Food? Item Selection and Price Differences Affect Low-Income Household Food Costs," Food and Rural Economics Division, Economic Research Service, U.S. Department of Agriculture, Agricultural Economic Report No. 759, http://www.ers.usda.gov/publications/aer759/AER759.PDF.

18. According to Matt Byers in his Alameda County Public Health Department report.

19. Oakland Unified School District, "Community Schools, Thriving Students: A Five Year Strategic Plan," Summary Report, Version 2.0, June 2011, http://www.communityschools.org/assets/1/AssetManager/Community-Schools-Thriving-Students-Strategic-Plan%20June%202011.pdf.

20. Oakland Unified School District served 7,261 breakfasts, 21,195 lunches, and 8,398 snack supplements in 2010–11, according to LeBarre. The federal government reimburses the district $2.77 for "free" lunches served to children from families at or below 130 percent of the federal poverty level; $2.37 for "reduced-price" lunches served to children from families between 130 and 180 percent of the federal poverty level; and $0.26 for paid lunches served. Reimbursement levels for breakfast and snacks are less than that. And money received from the federal government must cover the cost of salaries and facilities, as well as food. See Federal Register/Vol. 76, No. 139/Wednesday, July 20, 2011/Notices, p. 43256 for more information.

21. The Oakland feasibility study, conducted by the Center for Ecoliteracy, was funded by the TomKat Charitable Trust and S. D. Bechtel, Jr. Foundation. For more information, see http://www.ecoliteracy.org/downloads/rethinking-school-lunch-oakland-feasibility-study.

22. For more information on Father Boyle, see http://homeboy-industries.org/gregboyle.php.

CHAPTER NINE

1. From "Circle: A Place Where All Voices Are Equal," in *Feet to the Fire: The Rethinkers' Guide to Changing Your School* (Kids Rethink New Orleans Schools, 2011, http://www.therethinkers.com/spring-2011/).
2. Reggio Emilia is a model of early childhood education developed in Italy. It views the child as an active participant in learning; the physical environment of the school as key to stimulating learning; teachers, parents, and students as collaborators in learning; and the importance of making learning visible through the display of student work. For more information, visit http://zerosei.comune.re.it/inter/index.htm.

CHAPTER TEN

1. Andrew Revkin, "Google Earth Fills Its Watery Gaps," *New York Times,* February 2, 2009, http://www.nytimes.com/2009/02/03/science/earth/03oceans.html?pagewanted=all.
2. David Biello, "Ocean Impact Map Reveals Human Reach Global," *Scientific American*, February 15, 2008, http://www.scientificamerican.com/article.cfm?id=ocean-impact-map.
3. National Center for Ecological Analysis and Synthesis, "Data: Impacts," http://www.nceas.ucsb.edu/globalmarine/impacts.

RESOURCES

INTRODUCTION

From Breakdown to Breakthrough

Collaborative for Academic, Social, and Emotional Learning
Chicago, IL
(312) 226-3770
www.casel.org

State Education and Environment Roundtable
Poway, CA
(858) 676-0272
www.seer.org

Stockholm Environment Institute
Stockholm, Sweden
+46 8 674 70 70
www.sei-international.org

PART ONE: STANDING STRONG ON A COAL MOUNTAIN

Chapter 1: Lessons from a Coal Miner's Daughter

Center for Climate and Energy Solutions (formerly Pew Center on Global Climate Change)
Arlington, VA
(703) 516-4146
www.pewclimate.org

International Energy Agency
Paris, France
+33 1 40 57 65 00
www.iea.org/

Kentuckians for the Commonwealth
London, KY
(606) 878-2161
www.kftc.org

National Energy Foundation
Salt Lake City, UT
(800) 616-8326
www.nef1.org

Rainforest Action Network
San Francisco, CA
(415) 398-4404
www.ran.org

U.S. Environmental Protection Agency, Environmental Education
Washington, DC
(202) 564-0443
www.epa.gov/enviroed

Chapter 2: Taking a Power Trip

Center for Health and the Global Environment
Harvard Medical School
Boston, MA
(617) 384-8530
chge.med.harvard.edu

Council of International Educational
Exchange
Portland, ME
(207) 553-4000
www.ciee.org

Education Network for Global and
Grassroots Exchange (ENGAGE)
Barron, WI
info@engagetheworld.org
globalgrassroots.wordpress.com

iLoveMountains.org
Boone, NC
(877) 277-8642
The Institute for Southern Studies
Durham, NC
(919) 419-8311
www.southernstudies.org

The Nature Conservancy
Arlington, VA
(703) 841-5300
www.nature.org

Sierra Club
San Francisco, CA
(415) 977-5500
www.sierraclub.org

Southern Energy Network
Student networks in multiple states
info@climateaction.net
www.climateaction.net

SouthWings
Asheville, NC
(828) 225-5949
www.southwings.org

University of Maryland Center for
Environmental Science
Cambridge, MD
(410) 228-9250
www.umces.edu

PART TWO: FROM ANGER TO ACTION IN OIL COUNTRY

Chapter 3: The Heart of the Caribou

The Goldman Environmental Prize
San Francisco, CA
(415) 249-5800
www.goldmanprize.org

Public Media Center
San Francisco, CA
(415) 434-1403
www.publicmediacenter.org

U.S. Energy Information Administration
Washington, DC
InfoCtr@eia.gov
www.eia.gov

U.S. Fish and Wildlife Service
Washington, DC
(800) 344-9453
www.fws.gov

Chapter 4: Beyond Whining

Rethinkers: Kids Rethink New Orleans
Schools
New Orleans, LA
(504) 208-2813
www.therethinkers.com

Social Entrepreneurs of New Orleans
New Orleans, LA
(504) 656-4581
www.seno-nola.org

Twomey Center for Peace through
Justice, Loyola University
New Orleans, LA
(504) 864-7433
www.loyno.edu/twomey/

PART THREE: SHARED WATER: MOVING BEYOND BOUNDARIES

Chapter 5: Water Wars and Peace

Eawag: Swiss Federal Institute of Aquatic Research
Dübendorf, Switzerland
+41 (0)58 765 55 11
www.eawag.ch

International Water Management Institute
Colombo, Sri Lanka
+94 11 2880000, 2784080
www.iwmi.cgiar.org

National Environmental Services Center
West Virginia University
Morgantown, WV
(800) 624-8301
www.nesc.wvu.edu

National Research Council
Washington, DC
(202) 334-2000
www.nationalacademies.org/nrc/

United Nations Educational, Scientific and Cultural Organization (UNESCO)
Paris, France
+33 (0)1 45 68 10 00
www.unesco.org

United Nations Environment Programme
Nairobi, Kenya
(254-20) 7621234
www.unep.org

USDA Economic Research Service
Washington, DC
www.ers.usda.gov

Woodrow Wilson International Center for Scholars
Washington, DC
(202) 691-4000
www.wilsoncenter.org

World Health Organization (WHO)
Geneva, Switzerland
info@who.int
www.who.int

Chapter 6: From Restoration to Resilience

American Society of Landscape Architects
Washington, DC
(202) 898-2444
www.asla.org

AmeriCorps
Washington, DC
(202) 606-5000
www.americorps.gov

The Bay Institute
Novato, CA
(415) 878-2929
www.bay.org

Conservation International
Arlington, VA
(703) 341-2400
www.conservation.org

Environmental Working Group
Washington, DC
(202) 667-6982
www.ewg.org

International Rivers
Berkeley, CA
(510) 848-1155
www.internationalrivers.org

National Center for Ecological Analysis and Synthesis
Santa Barbara, CA

(805) 892-2500
www.nceas.ucsb.edu

National Geographic Society
Washington, DC
(800) 647-5463
www.nationalgeographic.com

National Oceanic and Atmospheric
Administration
Washington, DC
www.noaa.gov

PRBO Conservation Science
Petaluma, CA
(707) 781-2555
www.prbo.org

Smithsonian Institution National
Museum of Natural History
Washington, DC
(202) 633-1000
www.mnh.si.edu

Students and Teachers Restoring a
Watershed (STRAW)
(Part of PRBO Conservation Science)
(707) 781-2555 x358
www.prbo.org/cms/192

Virginia Institute of Marine Science
College of William & Mary
Gloucester Point, VA
(804) 684-7000
www.vims.edu

Woods Hole Oceanographic Institution
Woods Hole, MA
(508) 289-2252
www.whoi.edu

World Resources Institute
Washington, DC
(202) 729-7600
www.wri.org

**PART FOUR: NOURISHING
COMMUNITIES WITH FOOD**

**Chapter 7: Changing the Food
System, One Seed at a Time**

American Public Health Association
Washington, DC
(202) 777-2742
www.apha.org

Center for Agroecology & Sustainable
Food Systems
University of California, Santa Cruz
Santa Cruz, CA
(831) 459-3240
casfs.ucsc.edu

Center for a Livable Future
Johns Hopkins Bloomberg School of
Public Health
Baltimore, MD
(410) 502-7578
www.jhsph.edu/clf

Colonias Development Council
Las Cruces, NM
(575) 647-2744
www.colonias.org

Farm to Table
Santa Fe, NM
(505) 473-1004
www.farmtotablenm.org

La Semilla Food Center
Anthony, NM
(575) 642-2818
www.lasemillafoodcenter.org

New Mexico Collaboration to End
Hunger
(888) 363-6648
www.endnmhunger.org

New Mexico Environment Department
Santa Fe, NM
(800) 219-6157
www.nmenv.state.nm.us

Sierra Vista Growers
Anthony, NM
(575) 874-2415
www.sierravistagrowers.net

U.S. Department of Agriculture
Washington, DC

(202) 720-2791
www.usda.gov

U.S. National Park Service
Washington, DC
(202) 208-3818
www.nps.gov

W. K. Kellogg Foundation
Battle Creek, MI
(269) 968-1611
www.wkkf.org

Chapter 8: Forging the Food Justice Path

Alameda County Public Health
Department
Oakland, CA
(510) 267-8000
www.acphd.org

The California Endowment
Los Angeles, CA
(800) 449-4149
www.calendow.org

Harlem Children's Zone
New York, NY
(212) 360-3255
www.hcz.org

HOPE Collaborative
Oakland, CA
(510) 444-4133
www.hopecollaborative.net

Kaiser Permanente
Oakland, CA
healthy.kaiserpermanente.org

Oakland Unified School District
Oakland, CA
(510) 879-8200
www.ousd.k12.ca.us

People's Grocery
Oakland, CA
(510) 652-7607
www.peoplesgrocery.org

Revolution Foods
Oakland, CA

info@revolutionfoods.com
www.revfoods.com

Urban Strategies Council
Oakland, CA
(510) 893-2404
www.urbanstrategies.org

PROFESSIONAL DEVELOPMENT STRATEGIES

Sustainable Table
New York, NY
(212) 991-1930
www.sustainabletable.org

RELATED ORGANIZATIONS

Food/Health

Action for Healthy Kids
Chicago, IL
(800) 416-5136
www.actionforhealthykids.org

Alliance for a Healthier Generation
Portland, OR
Info@HealthierGeneration.org
www.healthiergeneration.org

American School Health Association
Bethesda, MD
(301) 652-8072
www.ashaweb.org

Center for Science in the Public Interest
Washington, DC
(202) 332-9110
www.cspinet.org

The Edible Schoolyard Project
Berkeley, CA
(510) 843-3811
www.edibleschoolyard.org

Food Research and Action Center
Washington, DC
(202) 986-2200
www.frac.org

The Food Trust
Philadelphia, PA
(215) 575-0444
www.thefoodtrust.org

Healthy Schools Campaign
Chicago, IL
(312) 419-1810
www.healthyschoolscampaign.org

Healthy Schools Network, Inc.
Washington, DC
(202) 543-7555
www.healthyschools.org

Kids Gardening
National Gardening Association
South Burlington, VT
(800) 538-7476
www.kidsgardening.org

Let's Move!
Washington, DC
www.letsmove.gov

National Farm to School Network (joint program)
Center for Food & Justice.
Los Angeles, CA
(323) 341-5095
Community Food Security Coalition
Portland, OR
(503) 954–2970
www.farmtoschool.org

Partnership for a Healthier America
Washington, DC
(202) 842-9001
www.ahealthieramerica.org

School Nutrition Association
National Harbor, MD
(301) 686-3100
www.schoolnutrition.org

Slow Food USA
Brooklyn, NY
(718) 260-8000
www.slowfoodusa.org

Society for Nutrition Education and Behavior

Indianapolis, IN
(800) 235-6690
www.sneb.org

Environment

Alliance to Save Energy
Washington, DC
(202) 857-0666
www.ase.org

Center for Biological Diversity
Tucson, AZ
(866) 357-3349
www.biologicaldiversity.org

Center for Environmental Health
Oakland, CA
(510) 655-3900
www.ceh.org

Center for Whole Communities
Fayston, VT
(802) 496-5690
www.wholecommunities.org

Children & Nature Network
Santa Fe, NM
(505) 603-4607
www.childrenandnature.org

Children's Environmental Health Network
Washington, DC
(202) 543-4033
www.cehn.org

Clean Water Fund
Washington, DC
(202) 895-0432
www.cleanwaterfund.org

Earth Island Institute
Berkeley, CA
(510) 859-9100
www.earthisland.org

Earthwatch Institute
Boston, MA
(800) 776-0188
www.earthwatch.org

Environmental Literacy Council
Washington, DC
(202) 296-0390
www.enviroliteracy.org

Natural Resources Defense Council
New York, NY
(212) 727-2700
www.nrdc.org

Post Carbon Institute
Santa Rosa, CA
(707) 823-8700
www.postcarbon.org

U.S. Green Building Council
Leadership in Energy and Environmental
Design (LEED)
Washington, DC
(800) 795-1747
www.usgbc.org

The Water Project
Concord, NH
(800) 460-8974
www.thewaterproject.org

World Wildlife Fund
Washington, DC
(202) 293-4800
www.worldwildlife.org

Worldwatch Institute
Washington, DC
(202) 452-1999
www.worldwatch.org

Education

Center for Place-Based Education
Antioch New England Institute
Keene, NH
(603) 283-2105
www.antiochne.edu/anei/cpbe

Go Green Database
Edutopia
San Rafael, CA
(415) 662-1600
www.edutopia.org/go-green

Green Charter Schools Network
Madison, WI
(507) 313-6273
www.greencharterschools.org

Green Schools Alliance
New York, NY
info@greenschoolsalliance.org
www.greenschoolsalliance.org

Green Schools Initiative
Berkeley, CA
(510) 525-1026
www.greenschools.net

Institute for Humane Education
Surry, ME
(207) 667-1025
www.humaneeducation.org

National Environmental Education
Foundation
Washington, DC
(202) 833-2933
www.neefusa.org

North American Association for
Environmental Education
Washington, DC
(202) 419-0412
www.naaee.net

U.S. Partnership for Education for
Sustainable Development
Washington, DC
uspesd@gmail.com
www.uspartnership.org

INDEX

A

Academic achievement, and ecological well-being, 8–9

Action for Healthy Kids, 101

Agriculture: industrial, 83, 91; and water consumption, 91

Alaimo, K., 101

Alaska Community Database Community Information Summaries, 49

Alaska Native Claims Settlement Act (1971), 50

Alaskan oil, 47–48

Alavian, V., 72

Allen, E., 77

American Public Health Association, 91

American Society of Landscape Architects, 83

"America's Most Endangered Mountains," 40

Anderson, R., 53

Annan, K., 66

ARAMARK, 55, 62

Arciniega, R., 105

Arctic National Wildlife Refuge (ANWR), 45–46, 54; drilling, prospect of, 50–51

Arctic Village, 46

B

Bachemin, A., 58–59

Bacon, F., 68

Baker, Jo, 103

Balow, Z., 6

Barnaby, W., 72

Barrouquere, B., 27

Barry, D., 25

Bass, A., 103

Bass, R., 49

Belt, D., 70

Benayas, J.M.R., 83–84

"Benefits of SEL" (Collaborative for Academic, Social, and Emotional Learning), 9

Bennett, L., 56, 61

Berkeley Unified School District, 6

Berntell, A., 66, 70

Berry Center, 91

Berry, W., 9, 23, 26, 27, 27–28, 29, 33, 36, 91, 124, 133–134

Beshear, S., 23, 27–28

Beyond Whining, 55

Biello, D., 82, 83, 127

Big Thirst: The Secret Life and Turbulent Future of Water (Fishman), 67

Blanton, T., 9, 23–24, 26, 27, 28, 29–32; Canary Project, founding of, 32; Kentuckians for the Commonwealth, 32; lessons learned about effective leadership, 32

Block, B., 127

Blue Gold: World Water Wars (documentary), 66

Bohrer, B., 47

Borlaug, N., 91

Borrell, B., 83

Boutros, B., 66

Boyle, G. (Rev.), 109

Brockovich, E., 24

Broenstein, S., 59

Brooks Range, 45

Buckley, P., 6

Burkes, B., 60, 61, 62

Bush, G. H. W., 53

Byers, M., 99, 104

C

California Endowment, 109

California State Adopt-A-Species Program, 80

Canada, G., 102

Capra, F., 3, 6–7, 135

Caribou people, 45–46, 48–49

Cariou Rising (Bass), 49

Carter, J., 47, 50

THE CENTER FOR ECOLITERACY

The Center for Ecoliteracy is a nonprofit foundation dedicated to education for sustainable living. Known for its pioneering work with school lunches, gardens, and integrating sustainability into K–12 curricula, the Center has worked with educators from across the United States and six continents. Through its Smart by Nature initiative, the Center supports educators advancing sustainability in food practices, building and resource use, community connections, and teaching and learning. Through its Rethinking School Lunch initiative, the Center promotes the essential connection between healthy school food and teaching about food systems and choices. The Center's services include publications, seminars, academic program audits, coaching for teaching and learning, in-depth curriculum development, keynote presentations, technical assistance, and a leadership training academy.

Center resources include the books *Smart by Nature: Schooling for Sustainability*, which showcases inspiring stories about school communities across the nation; *Ecological Literacy: Educating Our Children for a Sustainable World*; a conceptual framework for integrating learning in K–12 education (*Big Ideas: Linking Food, Culture, Health, and the Environment*); and the cookbook and professional development guide, *Cooking with California Food in K–12 Schools*. The Center's website offers hundreds of downloadable resource materials, including lessons and activities, an extensive planning framework (the *Rethinking School Lunch Guide*); the *Rethinking School Lunch Oakland Feasibility Study*; discussion guides for films such as *Food, Inc., Nourish: Food + Community*, and *The Last Mountain*; and essays by leading writers and experts.

The Center for Ecoliteracy was cofounded by Fritjof Capra, physicist and systems thinker; Peter Buckley, former CEO of Esprit International and environmental philanthropist; and Zenobia Barlow, now its executive director. It is located in the award-winning David Brower Center, a home for environmental and social action in Berkeley, California.

THE AUTHORS

Daniel Goleman's *Emotional Intelligence: Why It Can Matter More Than IQ* (Bantam, 2005) was on the *New York Times* bestseller list for a year and a half, with more than five million copies in print in forty languages. He has written books on topics including self-deception, creativity, meditation, social and emotional learning, and the ecological crisis. He is a cofounder of the Collaborative for Academic, Social, and Emotional Learning (CASEL), he codirects the Consortium for Research on Emotional Intelligence in Organizations, and he is a board member of the Mind & Life Institute. He won a Lifetime Career Award from the American Psychological Association and is a Fellow of the American Association for the Advancement of Science.

Lisa Bennett is the communications director for the Center for Ecoliteracy and a contributor to books including *The Compassionate Instinct: The Science of Human Goodness* (W. W. Norton, 2010), *Smart by Nature: Schooling for Sustainability* (Watershed Media, 2009), and *A Place at the Table: Struggles for Equality in America* (Oxford, 2001). A longtime writer, her articles have appeared in the *Christian Science Monitor, Chronicle of Higher Education, Education Week, Greater Good, Harvard Review, More, Mothering, New York Times, Newsday,* and elsewhere. She currently blogs at *The Huffington Post* and is a former Fellow at Harvard University's Center on the Press, Politics, and Public Policy in the John F. Kennedy School of Government.

Zenobia Barlow is executive director of the Center for Ecoliteracy. A pioneer in creating models of schooling for sustainability, she has designed strategies for applying ecological and indigenous understanding in K–12 education, including the Food Systems Project and the Rethinking School Lunch and Smart by Nature initiatives. She coedited *Ecological Literacy: Educating Our Children for a Sustainable World* (Sierra Club Books, 2005) and *Ecoliteracy: Mapping the*

1

Terrain (Learning in the Real World, 2000). She serves on the board of directors of the David Brower Center and is a Fellow of the Post Carbon Institute.

Carolie Sly is education program director at the Center for Ecoliteracy, where she is responsible for organizing and directing the Center's professional development work for educators. She founded a high school for at-risk youth and taught at San Francisco State University and public schools in Davis and Napa, California. Carolie earned a doctoral degree in science education from the University of California, Berkeley, and coauthored the award-winning *California State Environmental Education Guide* (1995) and the Center for Ecoliteracy's *Big Ideas: Linking Food, Culture, Health, and the Environment* and the discussion guides for the films *Food, Inc.* and *The Last Mountain*.